This Fragile Land

Inscribed for Carol
Schumann ~

With my very best
wishes!

Paul Johnsgard

Oct 11, 1995

This Fragile Land

A Natural History of the Nebraska Sandhills

Paul A. Johnsgard

University of Nebraska Press *Lincoln and London*

Publication of this volume was assisted by The Virginia Faulkner Fund,
established in memory of Virginia Faulkner, editor-in-chief of the
University of Nebraska Press.

⊗ The paper in this book meets the minimum requirements of
American National Standard for Information Sciences—Permanence
of Paper for Printed Library Materials, ANSI Z39.48-1984.

Library of Congress Cataloging in Publication Data
Johnsgard, Paul A.
This fragile land: a natural history of the Nebraska Sandhills /
Paul A. Johnsgard.
p.   cm.
Includes bibliographical references (p.   ) and indexes.
ISBN 0-8032-2578-4 (cl.: alk. paper)
1. Natural history—Nebraska—Sandhills.   2. Sand dune ecology—
Nebraska—Sandhills.   3. Sandhills (Neb.)   I. Title.
QH105.N2J64   1995
508.782—dc20
94-36409
CIP

For the people of the Sandhills, who have known and
respected this land far longer than I

Contents

Illustrations

Tables

Preface

People visiting Nebraska for the first time might, upon studying a highway map, scan it carefully without ever realizing that within the state lies the largest area of sand dunes in the Western Hemisphere, a region larger than Vermont and New Hampshire combined. Other than showing a notable scarcity of roads and towns between the Platte and Niobrara Rivers, tourist maps provide no clue as to what the land itself is like, beyond identifying two rather inexplicable "national forests" incongruously situated in what might be considered the heart of the Great Plains.

In 1961, the summer after I had moved to Nebraska to teach at Lincoln, my wife and I loaded our family of three preschool children into our Volkswagen microbus. We then set out on an exploratory trip to western Nebraska via U.S. Highway 2, hoping to see if that really is where the West begins. We initially camped near Thedford, in the Bessey Division of the Nebraska National Forest, and as I drove the tent stakes into a pure sand substrate, I began to realize that interior Nebraska is indeed sandy! That realization became even stronger the next day, as I turned south off the pavement at Lakeside onto an unnumbered sand road, hoping to drive the 30-odd miles to Crescent Lake National Wildlife Refuge. It was an oppressively hot June day, and our vehicle had no air conditioning and only very limited power for traversing the dunelike hills. I soon began to feel that we were alone in a desert, on a narrow and seemingly endless one-lane road without any other traffic or signs of human habitation except widely scattered windmills and seemingly lifeless ranch buildings. My fears of running out of gas or becoming stuck in the sand, without much water and with several small children to think about, finally caused me to turn around and head back toward Highway 2.

Later, I explored the farthest back roads of the Sandhills on my own, occasionally becoming stuck axle-deep in soft sand or completely lost on unmarked roads but eventually learning how to cope with this near-wilderness region. In the process I learned to appreciate the Sandhills' special plants and animals, the long, late afternoon shadows that softly caress the rounded hills, and the often spectacular, uncluttered summer sunsets. I finally began to realize that the true heart and spirit of Nebraska is not to be found in our increasingly stressful eastern cities, our vastly overrated university athletic programs, or even in the historic but now dying Platte River that whispers sad dirges of times past as it glides eastward to meet an equally altered and degraded Missouri. Rather, the state's pioneer spirit persists in the quiet recesses of our Sandhills, particularly in the fortitude of the people who once homesteaded them and whose descendants sometimes still live there.

This book represents a kind of love letter to the Nebraska Sandhills and especially to their inhabitants past and present, including people, plants, and animals. It is also an appreciation for the many wonderful adventures I have had there, if only as a seasonal visitor. Since 1978 I have spent part of almost every summer at the University of Nebraska's Cedar Point Biological Station, where from any north-facing window one can see the southern edge of the Sandhills simply by looking across Lake Ogallala. There, resembling the shoreline of a permanently frozen ocean, the Sandhills constantly beckon, Sirenlike, to both seasoned biologist and student, inviting a closer look but simultaneously proclaiming, "I am unique, I am fascinating, but if you are not careful you may regret it." In this relatively simple, distinctly rigorous ecosystem, biologists can often detect the effects of natural selection firsthand and can more readily evaluate ecological processes such as food chains and interspecies competition than is possible in more diverse habitats. Nevertheless, there are nearly a thousand species of vascular plants and vertebrate animals to be found in the Sandhills, to say nothing of the innumerable invertebrates, and each has a unique ecological story to tell.

This book tells only a few of those stories, but it tries to place the Sandhills in historic, geographic, and ecological perspective relative to the rest of our state. It avoids excessive tabular data, in-text citations, and undue use of technological terms and scientific names. However, I have provided a substantial number of appendices, including a glossary, a geologic timetable, lists of scientific names, and plant and animal checklists, so that persons searching for such information can locate it. There is also a bibliography of more than 200 sources. Readers interested only in the ecology of the present-day Sandhills and not in their geology or historical geography might start with Part 2 and perhaps later go back to survey Part 1.

During the academic year 1992–93 I was fortunate in having a teaching-free semester for intensive work on the book. I was also aided by a variety of biologist colleagues who offered practical advice, suggested relevant literature, and checked appropriate passages for accuracy. These people included several on the faculty of the Ecology and Organismal Biology section of the university's School of Biological Sciences. Members of the university's Geology Department also advised me, as did my geologist son Scott. I was aided especially by Bill Scharf, associate director of the Cedar Point Biological Station, with whom I regularly exchanged ideas. Several graduate and undergraduate students at Cedar Point also checked various chapters for readability, as did a few nonstudent friends. Two additional outside reviewers, David Armstrong and Gary Schnell, provided excellent advice. To all these people, and to any I may have overlooked, I offer my thanks. I must also acknowledge all the pleasures I have received from the wildlife I have been privileged to observe in the Sandhills over the past 15 years; I only hope I have done my subjects minimal justice.

Last, I thank the people of the Sandhills, who are also subjects of this book—if only secondary ones—and who invariably helped me when I most needed assistance. The most recent such Sandhiller was one who refused payment for extricating my car from a five-mile sand trap optimistically described by the Nebraska Highway Department as an "unimproved road." Instead, he simply asked me to stop and assist him the next time he might need it. He smiled and said, "You may not recognize me, because I may not look the same, but I think you will know me nonetheless." I believe that I shall.

PART ONE

Echoes of Eden

A Nebraska Overview

CHAPTER ONE
A Grassy Ocean
The Dunes of Ancient Yesterdays

Fig. 1. Adult male pronghorn.

THE FACE OF NEBRASKA is sandy. From the Platte Valley to the Niobrara, and from the Panhandle to the Missouri Valley, one need hardly scratch the state's surface to find sand. A veritable sand sea covers roughly a quarter of the state, encompassing in total about 50,000 square kilometers and stretching a maximum of some 425 kilometers along an east-west axis and up to 210 kilometers in a north-south direction. In fact, the Sandhills resemble a gigantic eastward-pointing handprint that has been firmly pressed down on the state. The broad wrist begins in the Panhandle's eastern Box Butte and Morrill Counties (see figure 48, with Appendix 5, for a map of counties); a short thumb protrudes southeastward into Dawson County toward Cozad; and a series of tapering fingers extends two-thirds of the way across the state to Holt, Antelope, Boone, and Greeley Counties. Within this vast region sand dunes stretch gracefully and endlessly from horizon to horizon. In height they range up to nearly 140 meters above their intervening valleys, and their gritty surfaces are clothed in a fragile mantle of grassy vegetation. The region is in fact the largest area of sand dunes in the Western Hemisphere, but in marked contrast to the other major sand seas of the world, it is almost entirely vegetated. The plants that cover it are a mixture of prairie species that somehow have managed to solve the problems of surviving on an utterly unstable substrate, in addition to the stresses of living in a climate that ranges from bitterly cold and wholly cheerless in winter to searingly hot and dry in late summer. It is, in short, not a place for the unprepared.

If one not only scratches the surface of a dune in the central Sandhills but also probes downward, it will become apparent that the Sandhills are built on a sandy floor, and this sandy floor in turn is constructed over a sandy basement. In some places it may be necessary to dig some 300 meters deep before encountering anything but sand or sandy gravel! If sand itself were truly valuable, Nebraska would be one of the richest states imaginable. Of course, this is not the case; one has to search elsewhere to find the true riches of the Sandhills. But one cannot hope to appreciate the present Sandhills without some understanding and comprehension of their past.

The story starts millions of years ago. Early in the Cenozoic era (see the geologic timetable in Appendix 1), the so-called age of mammals that began more than 60 million years ago, much or all of what is now Nebraska lay submerged in a vast inland sea, even as the Rocky Mountains and Black Hills were slowly rising to the west and north. During Oligocene times, between about 37 and 28 million years ago, there was a slowing rate of mountain-building and volcanic activity in the Rockies. During the same time there was also a general uplifting of the continental interior to above the ancient ocean level, the land area thus gradually expanding and forming what is now the central Great Plains. Materials accumulating to the east of the Rocky Mountains initially consisted mostly of wind-driven silt and volcanic ash. These deposits, such as the Brule and Chadron Formations of the White River Group, can still be seen where they are exposed locally, such as at the eroded base of Scotts Bluff. Probably the general climate of the times was dry, windy, and dusty when these mostly airborne materials found their way into what is now Nebraska. The Brule and Chadron Formations eventually reached a thickness of as much as 380 meters in present-day Cherry County, as the Great Plains began to resemble a gigantic plateau sloping gently eastward from the Colorado mountains. In the latter part of Oligocene and early Miocene times a second layer of materials began to overlie the earlier White River depositions. These consisted of similar wind-carried and water-deposited (alluvial) materials that were primarily centered in the western portion of the present-day Sandhills, where they often filled up old valleys with sandy accumulations.

During the latter half of the Miocene epoch, from about 19 to 5 million years ago, a major period of alluvial deposition occurred in Nebraska. These sand-dominated deposits, which collectively are called the Ogallala Group and constitute the heart of the present-day Ogallala or High Plains aquifer, gradually began to accumulate over a vast basin that represented much of what is present-day Nebraska (figure 2). Most of these deposits were of sand or sandy gravel, which was carried eastward by once enormous but now vanished river systems fed by the presumably snowcapped mountains of eastern Wyoming and Colorado. One of these river valleys may have followed a course roughly parallel to the northern edge of the present-day Sandhills; another may have passed through their center and a third through their southern portions—all converging eastward. With these rivers came a vast trove of sandy alluvium, which was deposited almost all the way from present-day Nebraska's southwestern limits to its northeastern boundaries. In some places in the present western Sandhills this sandy layer is almost as thick as the Brule Formation strata lying directly below it, and the Brule

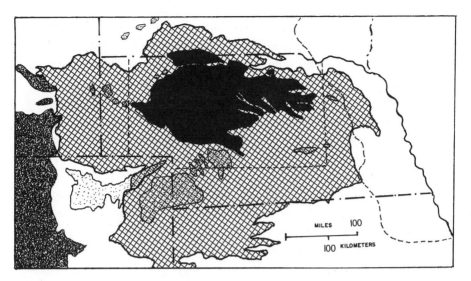

Fig. 2. Location of the Nebraska Sandhills (inked), major peripheral sandy areas (stippled), Ogallala aquifer (cross-hatched), Rocky Mountain piedmont (dark stippling), and limits of glaciation in Nebraska (dashed line). Dotted outline shows approximate coverage of Sandhills regional maps in this book.

Formation as it proceeds eastward gradually tapers out and disappears long before the Ogallala-related materials begin to diminish appreciably in thickness. It is this highly porous layer spread generously across Nebraska, thicker and thus richer in water-bearing sand than anywhere else in the Great Plains, that has produced the vast underground sponge known as the Ogallala aquifer. This tremendous reservoir of wonderfully fresh, pure water, which has been accumulating for millions of years, is the true hidden wealth of Nebraska, far more valuable to the state than the richest imaginable veins of Colorado gold could ever be, or the most diamond-rich beds to be found in South Africa.

During the latter part of the Miocene epoch the prevailing climate was fairly cool and dry, causing a prairielike vegetation to develop. With this kind of grassy mantle available, a wide variety of large grazing animals (rhinos, horselike chalicotheres, and others) appeared. Near the end of this epoch, about 10 million years ago, new volcanic activity in the western mountains occurred occasionally. These events brought in vast clouds of volcanic dust, which sometimes settled over the plains like a choking, deadly pall. An example of such an ashfall can be found just east of the present-day Sandhills, in Antelope County. The animals trapped in and smothered by this lethal cloud included single-toed horses, camels, giant hornless rhinos, prongbuck antelopes, and many others. One such deathtrap site, called the Poison Ivy Quarry, has recently been preserved and developed for public exhibition as Ashfall Fossil Beds State Historical Park.

During the subsequent Pliocene epoch, eastward-flowing rivers continued to dissect the western Great Plains, but little evidence of Pliocene deposits is to be found in present-day Nebraska. By the end of this epoch, however, some two million years ago, much of what is now Nebraska probably consisted of an undulating grass-covered plain, home to great herds of grazing animals (three-toed horses, zebras, camels, llamas, antelope) closely resembling the present-day temperate-climate ungulates of the Americas and Asia.

With the beginning of the Pleistocene epoch about two million years ago, a series of four continental glaciations and intervening warming or interglacial periods began. The first and second glaciations, called the Nebraskan and Kansan, had major geologic effects on what is now eastern Nebraska, especially by depositing massive amounts of glacial till. Their western limits of direct influence had slight impact on the Ogallala deposits and never reached the eastern edges of the present-day Sandhills. Nevertheless, the glaciers undoubtedly affected the climate of the entire plains area and brought with them from the north such famous Pleistocene mammals as mammoths and prehistoric bison, beginning about 1.5 million years ago. During the latter two major glaciations, the Illinoian and Wisconsinian, the associated ice sheets either barely reached the northeastern corner of present-day Nebraska or did not enter the region at all. Yet during that time snow-fed streams continued to flow out of the western mountains, augmenting the major river valleys of Nebraska with ever new layers of sand and silt. Additionally, during the last episode of glaciation, which occurred some 40 to 20 thousand years ago, windstorms of enormous magnitude must have swept across the plains. These winds sorted and sifted the lighter silty materials of the Ogallala and other exposed sediments, then carried the materials to various but often great distances before eventually depositing them in broad and diffuse bands stretching across eastern and southern Nebraska.

These wind-carried silts formed the original basis for the loessial soils now found mostly south of the Platte and along the eastern edges of the Sandhills, where they may reach thicknesses of as much as 20 meters, for example, in Custer County. Perhaps a good portion of these materials came from the surface layers of the developing Sandhills themselves, as was originally proposed by A. L. Lugn in 1935. The loess may have also have originated in or been supplemented from other regional sources, such as the floodplain of the Platte River.

There is no unanimity of opinion as to the age of the dunes themselves. Perhaps during that same general time the surface sands of the Sandhills became amalgamated and their surface topography differentiated, as has often been hypothesized in the past. More recently, however, geologists have been inclined to believe that the surface dunes of the Sandhills are of later formation. One hypothesis—based partly on radiocarbon dating of alluvial sands, soil profile development, and pond sediment analysis—suggests a dune development period that began no more than 8–10 thousand years ago, during a time of relative postglacial aridity. In this view, dry and windy climatic conditions resulted in massive melting of the continental ice sheet and widespread northward glacial retreat. There was a corresponding progressive expansion of prairie vegetation throughout the Great Plains, and associated northward retreat or restriction of boreal plants and animals to a few local refugia, such as the shaded north-facing slopes of the Niobrara Valley. Pollen remnants in interdune sediments of the northern Sandhills indicate that spruce trees were present in that general area about 12–13 thousand years ago, a time that would favor the presence of a cold-adapted vegetation in the northern Sandhills during the period of final glacial retreat.

During the generally arid and warm postglacial period 8–10 thousand years ago, the surface dunes of the Sandhills were probably largely unrestrained by vegetation. As a result, there may have been large-scale dune migrations across river valleys and other low-lying basins. One such movement may have occurred in what is now Keith County, where sand dunes evidently blocked the channel of the North Platte River. The dune dam eventually impounded a lake that was probably several times greater in volume than today's artificial Lake McConaughy and about twice its surface area. At its maximum size the lake extended well to the west of present-day Oshkosh (see figure 38). Sediments providing evidence of this now extinct lake, recently named Lake Diffendal by James

Swinehart, were discovered by Robert Diffendal Jr. of the University of Nebraska during the 1970s. The sand-based dam impounding Lake Diffendal must eventually have given way, allowing the North Platte to flow freely again until the present century, when engineers picked approximately the same site to construct Kingsley Dam and create Lake McConaughy.

Farther north in the Sandhills, similar dune dams probably crossed and blocked various creek and river valleys, altering and obstructing drainage patterns. According to the recent findings of James Swinehart and his colleagues, moving dunes probably blocked the Blue Creek drainage of Garden County and the Birdwood Creek drainage of Lincoln County between 3 and 11 thousand years ago. Rather than a single large impoundment, however, a series of lakes and wetlands probably developed behind these dunes, which, because of the "leaky" nature of the dams, allowed the blockages to persist rather than eventually to be swept away. At the northern end of the Blue Creek drainage system in Garden County there are nearly 200 small lakes, plus some 500 more in southern Sheridan County and about 20 in eastern Morrill County to the west. Altogether, the present Blue and Birdwood drainage systems have more than 1,000 associated wetlands within their collective basins, making this part of the Sandhills one of the most valuable areas in the entire state for wetland-dependent plants and animals.

Farther east, dune dams may also have formed in the North Loup and Calamus drainages. Evidence of ancient peat beds lying along the Calamus River indicates that standing-water wetlands must have once existed there, probably between 2 and 3 thousand years ago. The series of about 400 lakes and other wetlands currently lying beyond the headwaters of the Calamus River in Cherry County probably resulted at least in part from this same damming effect.

The wetlands in the upper Blue Creek drainage are of special interest because they occur over a broad area situated just west of an ancient subterranean uplift called the Chadron Arch, which extends southeastward from Sheridan to Hooker County, producing a closed drainage basin. Depending on the relative permeability of the substrate or perhaps upon the amount of finer-grained clays washing in from adjacent tablelands and accumulating in the bottoms, these wetlands have become variously sealed off and thus variably independent of the levels of the associated aquifer below and around them. As a result of their relatively impermeable substrates and occasional rainfall replenishments, they may not dry up as frequently as typical playa lakes (see Glossary), but their waters tend to accumulate minerals because of evaporation. Eventually, these sealed-off wetlands may become highly alkaline. This effect is especially evident in northern Garden and southern Sheridan Counties, where clay substrates—perhaps washed in from the adjoining High Plains to the west—have produced an array of extremely salty wetlands. These so-called "hyperalkaline" or "soda" lakes have far higher levels of dissolved solutes than any of those farther east in the Sandhills and have developed unique associated ecosystems of alkaline-tolerant plants and animals.

Over the past few thousand years the Sandhills have again become stabilized, mostly by "pioneering" types of sod-forming and bunchgrass vegetation. Their deep and interlocking root systems helped hold the sand in place until a thin soil of humus and other organic matter had accumulated, providing increased nourishment for subsequent generations of these plants and for additional, newly invading species. But evidence of dune building and prior dune movements persists today. The present-day shapes and orientations of these stabilized dunes, like fading snapshots of the gigantic waves that once raged over the surface of a now silent sea, provide mute testimony to the winds of an ancient past.

In a classic 1965 review of dune topography in the Sandhills, H. T. V. Smith suggested that this region underwent three major episodes of dune building, the first of which he thought might have occurred as early as the initial Wisconsinian glaciation, or at least 50 thousand years ago. On the basis of radiocarbon dating studies and other high-technology criteria, geologists such as James Swinehart placed the beginning of dune building much more recently. Judging from their evidence dune-building has continued periodically right down to less than a thousand years ago, probably during dryer and warmer climatic cycles.

Regardless of the actual ages of the present-day dunes, Smith believed that the first series of developments consisted of the formation of the large and generally very tall transverse dunes that formed as long ridges at right angles to the then prevailing northwesterly winds. These dunes now extend in parallel east-west ridges that occur over much of the northern and western Sandhills and are the highest and longest dunes in the entire region. This initial dune-building phase may have been followed by a period of greater precipitation, allowing for the gradual stabilization of these enormous dunes by mantling vegetation and by the development of new drainage patterns. The second phase evidently resulted in the formation of smaller dunes ranging up to about three kilometers in length, varying from 90 to 180 meters in width and from 10 to 30 meters high. Rather than being oriented at right angles to the prevailing northwesterly winds, these dunes are generally aligned in a northwest to southeast direction and are concentrated in the southeastern parts of the Sandhills. Smith believed that they may have formed during a period of semiarid interglacial climatic conditions. The last dune-building episode resulted in yet smaller dunes, in Smith's scenario. These often tended to be superimposed on the larger and older dunes that were partially reactivated by blowouts (see Glossary), thus variously reshaping their surfaces. These and the even smaller dunes that developed near the boundaries of the Sandhills have little or no clear orientation with respect to the winds, and they are often so low as to be nearly shapeless, owing to the limited amounts of available sand.

When dunes are consistently exposed to winds from a single directional sector, they tend to assume various recognizable shapes, depending on environmental variables such as sand volume, substrate configuration, and wind velocity (figure 3). The large wavelike, parallel-ridged dunes typical of the central Sandhills tend to take on a distinctly sinuous rather than strictly linear shape, as their leading edges move at somewhat differing rates. This results in a complex "barchanoid-ridge" pattern, which is the commonest of all the dune forms occurring in the Sandhills; this general type now occupies about a quarter of the total surface area of the Sandhills region. These great sand ridges, or "megadunes" may be as long as 40 kilometers in the western Sandhills of Cherry and Grant Counties but are often no more than about 8 kilometers in length. They may also reach great heights above their intervening valleys, 135 meters being the maximum recorded in the Sandhills, in northern Grant County (figure 4).

Barchanoid-ridge dunes grade into simpler "barchans," the classic crescent-shaped dunes that have paired points extending downwind. The concave outline of the barchan dune is formed by its slip face, the relatively steep slope that lies downwind of the dune's crest, where sand particles descend and accumulate after having been blown over its top. Some particles are also blown around to its side, thus forming the

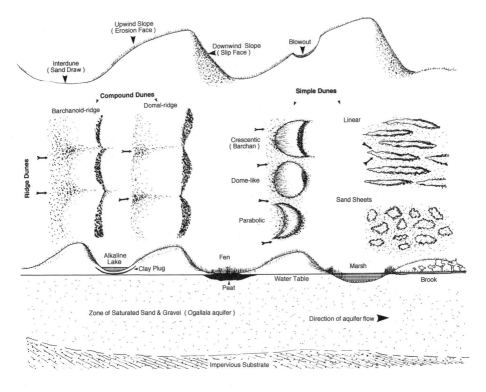

Fig. 3. Top: profile view of typical dune features (vertical scale exaggerated); *middle:* major Sandhills dune types as viewed from above; *bottom:* profile view of various surface and subsurface features in the Sandhills. Arrows indicate directions of prevailing winds.

graceful "horns" of the crescent. Typical barchans make up about 20 percent of the total surface area of the Sandhills and are best developed in northeastern Garden and southeastern Sheridan Counties.

Sometimes the pointed tips of crescent-shaped dunes are oriented toward, not away from, the prevailing winds. Such dunes can probably form only where enough vegetation is present to interrupt winds and thus influence and retard sand flow. In the Sandhills these so-called "parabolic" dunes are fairly rare, covering less than 5 percent of the total region. They are limited to the southwestern portions, from Morrill to Keith County, where they exist as isolated dunes or as parts of larger compound dunes. In Morrill County they have been found to average about 18 meters high and nearly 450 meters in length, alternating with

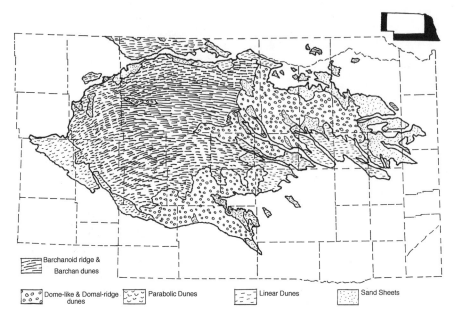

Fig. 4. Distribution of major dune types in the Sandhills region. Adapted from a map by J. Swinehart (in Bleed and Flowerday 1989).

interdune valleys that make up nearly half of the total surface area. These dunes have tips typically oriented northwesterly, and probably have resulted from long-ago winds out of the north to northwest, which is still the prevailing direction of winter winds in Nebraska (summer winds are more southerly).

Over much of the north central and south central Sandhills and covering about 20 percent of the total area are the generally large circular to elliptical dunes called "domelike." The simplest of these are widely spaced elliptical dunes about 25 meters high and averaging 760 to 1,050 meters in length and width. Larger and taller dunes of this general shape are more common and may consist of so-called compound and complex dunes representing multiple dune-forming processes, including those in which various dune types have been superimposed on a different original dune structure. Thus, elongated "domal-ridge" dunes, with collective lengths of 32 kilometers or more, occur in some parts of eastern Cherry and Brown Counties.

"Linear" dunes, with ridges that roughly parallel the mean direction of the prevailing winds, occupy about 10 percent of the Sandhills region. They are especially evident along the southeastern borders, from northwestern Custer and southern Blaine Counties east to Wheeler County. They average about 1.5 kilometers in length, 15 meters in height, and 150 meters in width, with the interdune valleys representing about 20 percent of the total surface area. Their mode of formation is still somewhat controversial, but most probably they were formed by winds coming from two distinctly different directions within a broad directional sector, producing two slip faces. In some Sandhills areas linear dunes have been formed on top of basically domelike dunes.

In areas of limited sand availability, irregular "sand sheets" of low to moderate relief (ranging up to 20 meters high) occur. These features cover about 15 percent of the total Sandhills region. They are especially common along the easternmost edges of the Sandhills in Rock, Holt, Wheeler, and Antelope Counties and also occur on the outlying sandy islands beyond the limits of the contiguous Sandhills. They have no consistent shape and probably are greatly influenced by the characteristics of the local substrate topography. Thus, at their extreme edges the Sandhills loose their unique identity and cannot be distinguished superficially from other kinds of landscapes.

Yet whatever their surface configurations, the Nebraska Sandhills are a very special place that has obviously been patiently shaped by water, wind, and time. One is well advised not to try to analyze the probable type of every dune or to identify the species of every plant. Rather, visiting the Sandhills should always partly be an exercise in visual aesthetics. What is present today is a fragile land, a land of no straight lines, where wind becomes artist and sand has metamorphosed into art. The Sandhills are a place of endless grass and countless hills, of unbroken horizons and broken hearts, of astonishing beauty and unimagined hardships, of abundant life and unexpected death. They are a region to be visited and revisited, to be long remembered and forever treasured.

CHAPTER TWO
Northern Shores
The Niobrara Valley

Fig. 5. American tree sparrow on willow in winter.

THE RELICS OF GLACIERS still haunt us. They emerge as bones and tusks of long-extinct elephants from the roadcuts and riverbanks of nearly every county in Nebraska. They line the hallways of countless museums as fragments and reconstructed skeletons of mammoths, mastodons, camels, antelopes, horses, and dozens of other animals that made the central Great Plains their home no more than 20–30 thousand years ago. They intrude on our daydreams of what the Great Plains must have been like when the first human hunting parties entered the region after crossing over the Pleistocene land bridge that once connected Asia with North America.

Few relics of the Pleistocene epoch are readily visible in the Sandhills today. As noted in chapter 1, pollen analysis from Minnechaduza Creek in the northern Sandhills suggests that spruce forests may have existed there about 12,600 years ago. Indeed, spruce trees evidently existed as far south as northeastern Kansas until about the same period and were probably common in western Iowa until about 11 thousand years ago. The few scattered groves of old ponderosa pines at the southern and eastern edges of the Sandhills lead one to believe that coniferous woodland was once more widespread in central Nebraska than it is today. More obviously, not far to the east of the Sandhills boundary there are deep layers of glacial till, glacier-carried boulders from farther north, ice-wedge casts (resulting from the substrate-contraction effects of permafrost), and other glacial features that are still firmly impressed upon the surface landscape of eastern Nebraska. Also in the northern Sandhills are numerous fens with boglike peat accumulations and an assortment of boreal plants such as cottongrass, probably representing late-Pleistocene relics that have survived within the state only in this cool and ecologically unique environment.

The Niobrara has its inconspicuous headwaters at an elevation of about 1,500 meters within an area known as the Hartville Uplift, in Niobrara County, Wyoming. It remains a rather scenically unimpressive river in its leisurely passage over the Panhandle region of western Nebraska. The Niobrara Valley then cuts directly through the northernmost edge of the western Sandhills. There it splits off a segment of the Sandhills that crosses the Nebraska border and extends a short distance into South Dakota, mostly in southern Bennett County. Between these two Sandhills sections the Niobrara River runs freely, eventually cutting down to a bedrock channel near Valentine. The bedrock layer, geologically called the Rosebud Formation, consists of late Oligocene, early Miocene strata of the Arikaree Group. The Niobrara is thus the only Nebraska river that flows directly over its underlying bedrock substrate; consequently, it is clearer and often more varied in velocity—with frequent stretches of rapids and riffles—than the numerous other Nebraska rivers, which have muddy or sandy substrates. Eventually, it merges with the Missouri River, at a point in Knox County that is now impounded by Gavins Point Dam. At the peak of the Kansan glaciation the western boundaries of the associated ice sheets evidently reached as far as 160 kilometers east of Valentine—approximately where the Niobrara River now joins the Missouri, in north central Knox County. Thus the Niobrara ends at a point only some 550 kilometers directly east of its Wyoming headwaters and about 1,000 meters lower in elevation, descending at a rather leisurely average rate of only two meters per kilometer.

By the time the Niobrara River reaches Cherry County, its associated valley is already considerable. There, its south-facing slopes locally expose much of the Miocene-age Valentine Formation (about 10–12 million years old, and up to 100 meters thick) and the less visible and younger Ash Hollow Formation (about 5–10 million years old), which lies directly above it. Beginning in the vicinity of Valentine and especially conspicuous farther to the east in Rock County, underground water originating in the Ogallala aquifer underlying the nearby northern Sandhills seeps along the top of the relatively impervious Rosebud Formation. The water eventually finds its way to small secondary canyons and valleys locally known as "springbranch canyons." There, along shady north-facing valley slopes protected from the hot summer winds, water emerges as permanently flowing cool springs and small waterfalls. (On the south-facing slopes across the river no comparable development of springs exists, because the associated Valentine and Ash Hollow Formations are of a loamy and silty nature that does not retain enough water to generate a flowing aquifer.) In this special microenvironment the combination of steep, shady canyon walls and the cool waters of the north-facing slopes has preserved a few biological remnants of late glacial and early postglacial regional history.

The retreat of the glacier and the associated boreal forest must have meant extinction or at least local extirpation for many cold-adapted species, but others were fortunate enough to find refuge in such cooler-than-normal sites as the Niobrara Valley. A detailed biogeographic anal-

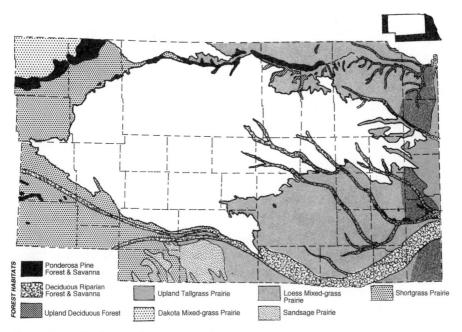

Fig.6. Generalized presettlement natural vegetation types in areas surrounding the Sandhills region. Adapted from map in Kaul 1975.

ysis of the plants and animals there, performed by Robert Kaul, Gail Kantak, and Steven Churchill, indicates that the valley's biota represents organisms of three quite distinct geographic affinities. First are the western-affinity species whose ranges encompass the Rocky Mountains, Great Basin, and Pacific Northwest. Second are those generally boreal-oriented species with ranges currently centered in the Canadian grasslands and forests far to the north, although they sometimes also occur in the Pine Ridge (see chapter 3) or the Black Hills. Third are those species whose main ranges are mostly associated with the evergreen and deciduous forests now lying mainly to the north, northeast, east, and southeast of the Niobrara Valley (figure 6).

Species having essentially western affinities were found to include 7 mosses, 14 vascular plants, and 28 animal species. Those of eastern, western, and (usually) northern affinities include 19 mosses, 40 vascular plants, and 18 animals. Those with predominantly eastern affinities include 15 mosses, 62 vascular plants, and 26 animals. Of the breeding birds, five western species reach their eastern limits in the Niobrara Valley (black-billed magpie, western wood-pewee, lazuli bunting, black-headed grosbeak, and western tanager), whereas 16 bird species reach approximately their western breeding limits there. Among these are such typically eastern deciduous forest species as the whip-poor-will, eastern wood-pewee, eastern phoebe, wood thrush, scarlet tanager, black-and-white warbler, ovenbird, American redstart, and red-eyed vireo. Six northern-oriented species (brown creeper, common snipe, red-breasted nuthatch, clay-colored sparrow, and tree swallow) reach or approach the southern ends of their breeding ranges in the Niobrara Valley.

Boreal affinities are most evident in the mosses and vascular plants of the Niobrara Valley, whose cool microclimate has allowed for the persistence of many species in the forests of the springbranch canyons and shaded valley walls. Among these is the paper birch, which is found nowhere else in Nebraska and no closer than the Black Hills. Among birds, the brown creeper and red-breasted nuthatch also have Black Hills breeding populations but otherwise may regularly breed no closer than the coniferous forests of Minnesota and the Rocky Mountains of Wyoming. Similarly, the pearl dace, a small minnow, is almost entirely limited in Nebraska to cool and clear streams in the Niobrara Valley, otherwise occurring no closer than central and southeastern Minnesota. Various other fish such as the blacknose shiner and several species of dace, also clearly glacial relicts, are largely limited to the Niobrara and other cool and spring-fed streams of the nearby northern Sandhills.

More important than its role as a refugium for glacial relicts is the function served by the Niobrara River, like the Platte River to the south, as a major biological corridor connecting the deciduous forests to the east with the conifer-dominated forests to the west. Indeed, in the approximately 160-kilometer stretch of the Niobrara from Valentine eastward may be seen one of the most important transition areas, or "suture zones," connecting east- and west-affiliated species of plants and animals in the entire Great Plains (figure 7). As the analysis of Kaul, Kantak, and Churchill makes clear, the Niobrara Valley represents a major east-west migration corridor that has probably existed since early postglacial times. The modern valley forest has certainly developed since the last glaciation, when forest colonizers from the south and east moved varying distances upstream along the Missouri and Niobrara Rivers until they reached their individual limits of tolerance to the increasing

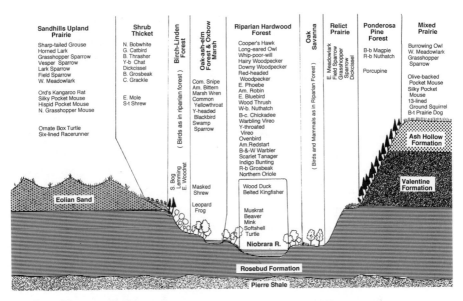

Fig. 7. Profile view of typical Niobrara Valley habitats near Valentine, and some associated vertebrates.

regional aridity. The species richness of trees, shrubs, woody and semi-woody vines, ferns, and mosses all show diminished numbers from east to west along this general drainage system. The reduction is especially evident among the ferns, which diminish from 33 to 6 species along the drainage gradient between the Missouri's mouth and the central Niobrara River. Similarly, trees and vines diminish to about a third of the maximum number present near the mouth of the Missouri River. Within the Niobrara drainage itself, however, the reductions in species richness of eastern-oriented plants are relatively minor; additionally, none of the western-oriented plants extends farther east than the mouth of the Niobrara. Among the mammals, the olive-backed pocket mouse is the only western form having its eastern limits in the valley. The eastern wood rat exists as an endemic subspecies in the valley, occurring well to the north of all other present-day populations of this southeastwardly oriented species.

Although east-west distribution gradient patterns of breeding birds are less dramatically apparent than those of plants, the central Niobrara Valley represents an extremely important east-west avian corridor for forest-dependent species. That is, it provides a meeting ground for the breeding ranges of several closely related species pairs of forest or woodland-edge birds whose common ancestors were probably separated into eastern and western components during Pleistocene times or even earlier. Their relatively recent "secondary contact" has occurred with the development of post-Pleistocene riverine forests, such as those of the Platte and Niobrara, across the otherwise relatively treeless central Great Plains. These birds include several eastern and western pairs of bird taxa that are currently considered by ornithological authorities to represent distinct species: eastern and western wood-pewees, scarlet and western tanagers, indigo and lazuli buntings, rose-breasted and black-headed grosbeaks. There are also eastern and western pairs currently regarded as only subspecies, such as the eastern (Baltimore) and western (Bullock's) races of the northern oriole, and the eastern (yellow-shafted) and western (red-shafted) races of the northern flicker.

The Niobrara Valley serves as a woodland corridor and mixing pot for all these previously separated forest-adapted or forest-edge population pairs. All of them potentially interact and at least most of them hybridize to varying degrees wherever they come into breeding contact in the riverine forests of Nebraska and South Dakota. Intergradation is especially the case for the subspecies pairs, such as the flickers and orioles, but also occurs at least occasionally between forms considered to be full species, such as the rose-breasted and black-headed grosbeaks. This sometimes results when two species' populations are so low at their extreme distributional boundaries that breeding opportunities (that is, available mating choices) may be quite limited: a bird of one species may take a mate of a closely related species out of necessity, and the potential for interspecific hybridization is thus set in motion.

This relatively rare situation provides a wonderful opportunity for biologists to study some of the most interesting aspects of biological evolution. These include the behavioral aspects of speciation and species maintenance: that is, the degree of development and the effectiveness of possible behavioral "isolating mechanisms" that may normally function to reduce or avoid hybridization in areas of interspecific contact. It may also be possible to investigate and test the effectiveness of the ecological isolating mechanisms that help maintain niche-related differences between such species. Additionally, the genetic consequences of hybridization and subsequent gene sharing can perhaps be studied for both populations when such hybridization is fairly frequent. Among the flickers the degree of hybridization is already so great that probably few if any "pures" of either type exist in the central Niobrara; on the other hand, the incidence of hybridization is apparently so low among the tanagers and wood-pewees as to be scarcely if at all detectable.

Evidence of previous hybridization interactions still apparent in the Niobrara Valley includes the presence of some apparently Pleistocene-related plant hybrid populations. These were evidently produced when there was a temporary Pleistocene contact between certain tree species that now allopatrically border the Great Plains grasslands. Thus, in the Niobrara Valley there is a surviving relict population of hybrid clones between the cottonwood and balsam poplar, and another between the quaking aspen and the eastern big-toothed aspen, even though the nearest currently existing populations of these species are about 300 kilometers apart.

Ponderosa pines extend eastward along the slopes of the Niobrara Valley to the central Niobrara, where they intermix with and eventually are replaced by eastern deciduous forest trees. In eastern Cherry County the pines are best developed on the south-facing slopes of the Valentine and Ash Hollow Formations and along the outcroppings of rock on the upper parts of north-facing canyons. Associated with this easternmost extension of the Pine Ridge forest are such coniferous forest mammals as porcupines and such birds as the black-billed magpie, the black-headed grosbeak, and the Bullock's race of the northern oriole. Red-breasted nuthatches often breed in the pines and probably reach their eastern range limits in the central Niobrara Valley where the ponderosa pines likewise disappear.

In the vicinity of Fort Niobrara National Wildlife Refuge and the Na-ture Conservancy's 45-kilometer-long Niobrara Valley Preserve, the sandhills community lies just to the south of the Niobrara canyon. The sand dunes there are just on the southern border of the Crookston Table, a high South Dakota benchland region of Tertiary origin that is characterized by sandy soils and aridity-adapted grasses and is transi-tional to the Sandhills themselves. This tableland is located in an area that originally supported mixed-grass prairie, but long-term grazing by introduced cattle, antelope, bison (on the Fort Niobrara refuge), and other large mammals has maintained it almost entirely as a shortgrass community dominated by buffalo grass and blue grama. Bison and elk were reintroduced into the Fort Niobrara area in 1925, and these once native species were supplemented in 1936 by longhorn cattle.

The most common bird species of the shortgrass tableland between the river canyon and the sandhill community to the south is the horned lark, followed closely by the lark sparrow and the western meadowlark. The western-oriented olive-backed pocket mouse reaches the eastern limits of its range in this community type. The rock-edged canyons leading to the Niobrara support breeding rock wrens and common poorwills, and reptiles such as prairie rattlesnakes, bull snakes, northern prairie lizards, and lesser earless lizards.

Shrub-dominated thickets occur between the zone of ponderosa pines at canyon edges and the deciduous forest community below, with fingerlike extensions of this shrubby community extending up many of the draws. Wild plum, chokecherry, and snowberry are among the commonest shrubs, and various forest-edge birds such as the yellow-breasted chat, gray catbird, northern bobwhite, and indigo bunting breed here. Lazuli buntings are quite rare in the central Niobrara Valley; they are outnumbered about 40-fold by the closely related and potentially hybridizing indigo buntings. Similarly, the yellow-shafted race of the northern flicker is about 20 times as common as the more westwardly typical red-shafted race; hybrids between these two types are common.

Remnant areas of tallgrass prairie can be found as small meadows on old river terraces and also surrounding cutoff channel (oxbow) wetlands. These prairie sites may support meadow jumping mice, short-tailed shrews, and eastern moles as well as such grassland breeding birds as vesper, lark, and field sparrows. In somewhat wetter sites, such as those around springs, swamplike communities dominated by willows, cattails, common reeds, and bulrushes occur, producing a typical wetland ecosystem with such animals as common watersnakes, northern cricket frogs, bull frogs, and northern leopard frogs. Frog-eating birds—American bitterns, green-backed herons, great blue herons—are often attracted to such sites, as are mink and raccoons. The northern-oriented swamp sparrow also nests in willow-rich wetlands, as probably also do common snipes, which are similarly northern breeders that only infrequently nest as far south as Nebraska.

The deciduous forest community, well developed along the canyon floor and riverside edges, includes American elm, green ash, boxelder, eastern cottonwood, ironwood, black walnut, and various willows. Bur oaks sometimes occur in combination with these and often dominate slightly dryer woodland sites on both sides of the river. The canyon's shaded northern slopes are more likely to include stands of paper birch and basswood. Here too may be found the typical avian representatives of eastern deciduous forests, such as the red-bellied woodpecker, whip-poor-will, wood thrush, red-eyed vireo, black-and-white warbler, oven-bird, and scarlet tanager. The more boreal brown creeper also nests here, the nests typically being placed behind the peeling bark of large dead trees, such as mature American elms that have been killed by Dutch elm disease.

The deciduous forest–associated ringneck snake occurs in these river-ine woods and reaches what is nearly its western Nebraska range limits in Rock County. Here too can be found eastern and western hog-nosed snakes, which search for toads (virtually their entire prey source) along sandy riverine forest floors and out into the nearby grasslands. Both hog-nosed snake species have special enlarged teeth on the sides of their rear upper jaws that enable them to puncture air-inflated toads, which employ this swelling device to avoid being swallowed by other snake predators.

For biologists, the Niobrara (over 100 kilometers of which is now part of the national Wild and Scenic Rivers system) is truly a river of dreams; its unique combination of natural beauty and biological interest is ade-quate to satisfy a biologist of any persuasion. Unlike the Platte, whose rich human history and biologic diversity stretches in a leisurely way across three states, the Niobrara compresses its primary focus of interest into a river distance of less than 160 kilometers, within which the eco-logical and evolutionary lessons it can teach are manifold and the aes-thetic rewards are equally pleasurable. As Charles Darwin unforgettably observed: "It is interesting to contemplate an entangled bank, clothed with plants of many kinds, with birds singing on the bushes, with various insects flitting about, and with worms crawling through the damp earth, and to reflect that these elaborately constructed forms, so different from each other, and dependent on each other in so complex a manner, have all been produced by laws acting around us."

CHAPTER THREE
Western Shores
The Pine Ridge and High Plains

Fig. 8. Male sharp-tailed grouse.

THE BLUFFS OF NEBRASKA are icons. They stand stalwart and resolute, like sentinels from the past, when they guided the explorers, fur trappers, and immigrants of an earlier America westward to new lands. Among them are Jail, Courthouse, and Castle Rocks and the more famous Chimney Rock and Scotts Bluff. All these prominent present-day landscape features are the highly eroded remnants of an ancient high plateau that formed during Oligocene and Miocene times, when materials were blown in and carried by streams from the mountains that were rising and were often volcanically active far to the west, mostly between 35 and 28 million years ago.

The Pine Ridge country of northwestern Nebraska extends from the Wyoming border northeastwardly to the Pine Ridge Indian Reservation along the South Dakota border. Rising in curved outline to the west and north of the Sandhills, its knifelike shape neatly severs the Sandhill's hand-shaped outline at the wrist. Such a blood-drenched image well reflects the historic tragedies that have been associated with the events at Wounded Knee, South Dakota, at the very tip of the Pine Ridge. A second, smaller escarpment of the same geologic age extends east from the Wyoming line not far from the North Platte River and continues along that valley toward Ogallala, where Castle Rock, Scotts Bluff, Chimney Rock, Courthouse Rock, and Jail Rock form a succession of easily recognizable landmarks along the Oregon Trail.

These distinctive escarpments—often sprinkled with trees, sometimes cloaked in fairly dense forests of ponderosa pine—usually rest on much earlier sedimentary strata. Their beds consist of Cretaceous-age shale or similar ancient materials dating from more than 65 million years ago, when the central plains lay below sea level or at most consisted of a low coastal plain. Nearly all the exposed faces of monumental bluffs such as Scotts Bluff and Castle Rock in the North Platte drainage and near Crawford and Chadron in the White River drainage are of Oligocene age. These materials consist of generally thick layers of grayish to pink clays, buff-colored silts or silty clays, and occasional thinner layers of whitish volcanic ash, the latter marking periodic episodes of major volcanic activity to the west. These multiple layers dating from the Oligocene epoch are collectively known as the Brule Formation of the White River Group, which is capped by the Gering Formation of the younger (Miocene) Arikaree Group. The Brule Formation alone is almost 150 meters thick in the Crawford area and nearly 200 meters thick

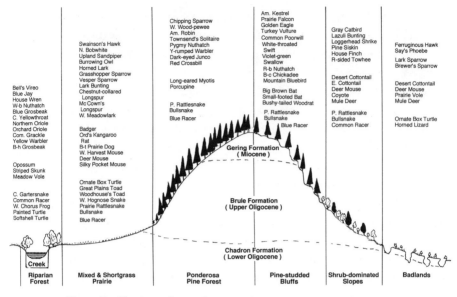

Fig. 9. Profile view of typical Pine Ridge and High Plains habitats, and some associated vertebrates.

around Scotts Bluff. These remnants leave the thoughtful observer to wonder what could possibly have happened to the incredibly massive amounts of now-missing clays and silts that were removed as these more resistant bluffs were gradually exposed through widespread landscape erosion (figure 9).

Except on the steepest slopes, the Pine Ridge escarpment is well vegetated. Its vegetational cover includes a distinctive assortment of natural communities that have many biological affinities with those of the Rocky Mountains and Black Hills to the west and north but grade into the high plains grasslands and Sandhills not far from their southern and eastern limits. Many of the plant and animal species of the Pine Ridge extend eastwardly along the slopes of the Niobrara Valley (as noted in chapter 2). Some of the birds that regularly breed in the Pine Ridge woodlands also filter down to spend fall and winter on the grasslands of the Sandhills or among the varied habitats of the North Platte Valley. For such reasons it is necessary to understand something of the Pine Ridge country if we are to appreciate the Sandhills fully.

The vegetational ecology of the Pine Ridge was studied in some detail by W. L. Tolstead in 1947. He estimated that ponderosa pines cover about 3,900 square kilometers in the Pine Ridge area of Nebraska, in addition to their more limited presence in the Wildcat Hills just south of the North Platte River and in the North Platte and Niobrara Valleys. Pines are usually absent from broad and gentle slopes with deep soils; these normally support prairie grasses, and pine seedlings evidently do not compete well under those conditions. Nor do they reproduce well in the deep shade of deciduous stands along creeks or in already established pine woodlands. On windswept hillsides and edges of grasslands the pines are usually under 10 meters tall and occur in densities of up to 500 trees per hectare. In areas where the trees are protected from winds, however, there may be as many as 5,000 trees per hectare, and they attain much greater heights, sometimes to 30 meters. Some trees in sites protected from logging are more than 250 years old. In dense groves the ground layer of vegetation is poorly developed and consists mostly of snowberry and poison ivy.

In canyons with small streams that provide a permanent water supply, deciduous forests develop. Usually quite limited in extent, these are dominated by green ash, American elm, and boxelder. Occasionally the ash will grow on protected slopes together with pines, and sometimes boxelder occurs alone in open stands. Hackberries also are present to a limited degree in these woodlands, sometimes reaching 20 meters in height, and in creek bottoms not subjected to frequent floods some fairly large willows and eastern cottonwoods often occur. The western-oriented mountain birch barely reaches into the Pine Ridge, in northern Sioux County.

In a later and more limited survey in Sheridan County, E. S. Nixon found the forests of ponderosa pine clearly associated with north-facing slopes and bottomlands. Such shrubby species as chokecherry, june-berry, and wild plum are also well represented. The chokecherries, june-berries, and pines increase relative to grassland flora on north-facing slopes; densities of grasses and other nonwoody plants decrease correspondingly. A few western redcedars sometimes occur among the pines as well, but they exist much more often as separate communities. Elsewhere in the Pine Ridge some other Rocky Mountain or boreal-oriented plants such as quaking aspen can be found, as well as fairly substantial stands of Rocky Mountain juniper. It is likely that the Pine Ridge escarpment has served as a limited refugium for some species that could not otherwise survive the occasional prairie fires of the high plains grasslands, or have benefited from the microclimatic variations brought about by the irregular topography of the escarpment.

On lower elevations, south-facing slopes, and open ridges a mixed-grass community is present, dominated by grama grasses and bluestems, with little bluestem especially characteristic. Other plants typical of mixed-grass prairie include sand bluestem, western wheatgrass, sideoats grama, and prairie sandreed grass. All these midgrass species extend eastward into the Sandhills grasslands.

The mammals and other terrestrial vertebrates of the Scotts Bluff National Monument area have been studied by M. K. Cox and W. L. Franklin, and their findings for this limited area can probably be applied to the Pine Ridge generally. Scotts Bluff rises to a height of about 180 meters above the surrounding High Plains terrain, and its crest is capped with a fairly dense stand of ponderosa pine and Rocky Mountain juniper. Shrub-dominated slopes of the monument, which covers about 1,200 hectares, exhibit mainly Rocky Mountain juniper and skunkbrush sumac. These shrubby areas grade into mixed-grass prairies of needle-and-thread, western wheatgrass, sideoats grama, prairie sandreed grass, and other grasses. Riverine woodlands adjacent to the North Platte River are dominated by eastern cottonwood, willows, boxelder, American elm, and green ash. Some riparian habitat in the monument is associated with an irrigation canal, and there is also an area of highly eroded and nearly vegetation-free "badlands" topography.

Although riverine woodland habitat constitutes only about 4 percent of the total Scotts Bluff National Monument area, Cox and Franklin found that it supported over half of the total 136 vertebrate species recorded, nearly 60 percent of the total bird species, and the greatest diversity of herpetofauna: 11 of a total of 12 species. The largest number of unique vertebrate species was associated with the deciduous riverine woodland community and the similar irrigation canal habitat. The permanent residents or potentially breeding migrants of these habitats included such woodland- and wetland-dependent species as the spiny softshell turtle, meadow vole, green-winged and blue-winged teal, wood duck, wild turkey, sora, red-headed and hairy and downy woodpeckers, yellow-billed cuckoo, bank swallow, eastern bluebird, house wren, blue jay, American crow, yellow warbler, common yellowthroat, warbling vireo, red-winged blackbird, and orchard oriole.

The richest mammalian habitat was found to be the mixed-grass prairie that makes up about half of the monument's total land area; it supported about 70 percent of the total mammalian species and nearly half of all the vertebrates recorded. Twelve vertebrate species were classified as unique to the mixed-grass prairie, including the Great Plains toad, upland sandpiper, burrowing owl, horned lark, grasshopper sparrow, black-tailed prairie dog, hispid pocket mouse, Ord's kangaroo rat, and badger.

The coniferous woodland habitat supported a relatively small representation of species for all three major vertebrate groups—collectively, less than 30 percent of the total. Only 3 of the 12 recorded reptiles and amphibians were present in the conifers, and only 8 of the 28 species of mammals. Of the 98 species of birds seen during the study, only 32 were associated with the coniferous habitat. The only vertebrate species that Cox and Franklin considered unique to the coniferous forest habitat on Scotts Bluff were the Townsend's solitaire (which has not yet been proved to breed locally) and the porcupine. Many woodland-breeding birds, however—including western wood-pewee, black-capped chickadee, American robin, rufous-sided towhee, and chipping sparrow—breed in the conifers as well as in the nearby dedicuous woodlands, and the robin reaches its greatest abundance there.

Farther north in the Panhandle, where the Pine Ridge coniferous forest is better developed, several bird species that otherwise occur only as close as the corresponding coniferous forests of the Black Hills probably breed fairly regularly. These include the pinyon jay, yellow-rumped warbler, solitary vireo, western tanager, pine siskin, red crossbill, and dark-eyed junco. More rarely nesting are other conifer-adapted species such as the pygmy nuthatch, western flycatcher, red-breasted nuthatch, Swainson's thrush, and Townsend's solitaire, for all of which very few Nebraska breeding records are known.

Several other species of birds are occasional to regular breeders on the higher bluffs and escarpments of the Pine Ridge, regardless of the actual vegetation; they seem to rely more on the physical isolation and outstanding visibility provided by the high bluffs than on the vegetation as such. These birds include the golden eagle, prairie falcon, and turkey vulture, which typically nest on cliff ledges and in recesses. White-throated swifts breed in small crevices in the bluff faces, as do nesting violet-green swallows, although the swallows are also likely to nest in old woodpecker holes in pines or other larger trees.

Lower escarpment outcrops or related erosion-dependent landscapes, including typical badlands topography, may attract nesting ferruginous hawks, great horned owls, common barn-owls, cliff swallows, rock wrens, and common poorwills. The Nebraska badlands, best developed in the Oglala National Grassland, support only low levels of birds and mammals, because they generally have very little vegetational cover and often no available surface water. The short-horned horned toad and the wandering gartersnake are limited to this part of the Nebraska Panhandle and are typical of this general habitat type.

Lying below the scenic escarpments of the Pine Ridge and spreading over their sunny south-facing slopes are the shortgrass plains, which reach north into western South Dakota, west toward Wyoming, and east to merge with the western Sandhills. The shallow and claylike soils here are high in minerals such as calcium. Surface runoff therefore occurs rapidly during the brief periods of spring and summer rainfall. As a result, shallow-rooted species of plants are favored, and most of the root systems are concentrated in the uppermost 30 centimeters or less of soil. The soil surface itself becomes extremely dry and hot during midsummer, causing the aboveground parts of all but the most drought-tolerant plants to wither. In this short-stature grassland the dominant species consist mostly of buffalo grass and various grama grasses, especially hairy grama, which often grow no higher than a half-meter above the surrounding plains. Unlike the bunchgrasses of the Sandhills, the dominant grasses of the shortgrass plains tend to form a continuous surface network of stems and leaves, especially where buffalo grass is more common than grama.

In somewhat sandier soils, taller grasses gradually gain ascendancy over the shortgrass species, and a shift toward the characteristic Sandhills vegetation may be detected. These midstature grasses include side-oats grama and red three-awn, as well as the leguminous silver-leaf scurfpea, which sometimes dominates acres of land with its silver-toned vegetation. A related species, wild alfalfa, grows in slightly wetter locations, forming relatively large and bushy shapes; these plants often break off at ground level at the end of the growing season to become one of the many kinds of "tumbleweeds" of the western plains.

On the shortgrass plains of eastern Wyoming, in the North Platte drainage of Goshen County and not far from the Nebraska border, M. H. Maxwell and L. N. Brown studied the ecological distribution of rodents in various community types (table 1). One of these was the

Table 1 Preferred habitat attributes of twelve High Plains grassland rodents

Species[a]	Soil Type[b]	% Bare Soil[b]	Ave. Plant Height (cm.)[b]	Community Type[c]	Capture Rate[d]	
Spotted ground squirrel	*sand*	>40	*50+*	sand dune	16.0	(41%)
Ord's kangaroo rat	*sand*	*>40*	*28–50*	sand dune	59.4	(62%)
Plains pocket mouse	sandy loam	>40	*50+*	sage–grass	7.1	(34%)
Prairie vole	sandy loam	<40	*50+*	sandreed–grama	25.5	(69%)
Western harvest mouse	*loamy sand*	<40	*50+*	yucca–grass	10.6	(38%)
Deer mouse	*loamy sand*	<40	*28–50*	yucca–grass	33.5	(32%)
Olive-backed pocket mouse	loamy sand	<40	2.5–25	grama–needlegrass	4.6	(100%)
13-lined ground squirrel	*loamy sand*	<40	2.5–25	grama–needlegrass	11.1	(43%)
Northern grasshopper mouse	*loamy sand*	<40	2.5–25	grama–needlegrass	21.3	(24%)
Hispid pocket mouse	*loam, sandy loam*	<40	2.5–25	grama–needlegrass– three-awn	5.4	(26%)
Silky pocket mouse	*loam*	>40	2.5–25	grama–needlegrass– three-awn	5.4	(56%)
Plains harvest mouse	loamy sand, loam	<40	2.5–25	sandreed–grama	10.2	(47%)

Source: Based on data of Maxwell and Brown (1968) from Goshen County, Wyoming, reorganized to show ecological variables apparently associated with highest trapping success.

[a]List is sequentially organized from species favoring sand substrates, tall vegetation, and more open habitats to those using loamy soils, short vegetation, and less open habitats.

[b]Italicized attributes are those that appear to be important variables favoring species occurrence, judging from relative rates of trapping success.

[c]Sand dune = variety of dune-adapted grasses and forbs; needlegrass = needle-and-thread; grama = blue grama; three-awn = purple three-awn; sandreed = prairie sandreed; sage = sand sagebrush and dwarf sagebrush.

[d]Capture rates (per 1,000 trap-nights) are shown for community types with highest trapping success; percentages (in parentheses) indicate proportion of all captures represented by that single community type.

buffalo grass–grama grass community, considered a high plains climax vegetation type on loam to sandy loam soils. On more sandy soils a typical Sandhills-like prairie sandreed–grama grass community occurs, and on sand dune areas a community made up of sand bluestem, prairie sandreed, leadplant, lemon scurfpea, and similar sand-tolerant species. Other community types present on the tablelands included a grama–needlegrass–three-awn community, and a grama–needlegrass community. Sage–grass and yucca–grass communities are best developed on bluff slopes and escarpments, and a more shrubby mountain mahogany–juniper–grass community on north-facing escarpment slopes. Virtually all the rodents in the area occur to some extent in two or more of these types. The total number of mammalian species captured per community ranged from 6 to 10 in grass-dominated communities and from 7 to 11 in the more mixed community types; the highest species diversity was observed in the sage–grass and yucca–grass communities.

In a somewhat similar study of small mammals on the High Plains in eastern Colorado, M. P. Moulton and others found that the effect of moderate grazing on small mammal populations was least in shortgrass communities. It was greatest in riparian woodlands, where mammalian diversity increased from four to eight species. A sand sagebrush community (similar botanically to the sage–grass community in the Wyoming study) had a slight increase in diversity under moderate grazing (from six to eight species), and the native shortgrass community had six species present under both lightly grazed and ungrazed conditions.

The most abundant species in the ungrazed sand sagebrush community was the western harvest mouse, which generally favors fairly tall vegetation; the northern grasshopper mouse was the most abundant in grazed and ungrazed shortgrass, and the deer mouse and western harvest mouse in grazed and ungrazed riparian woodlands. The ungrazed shortgrass community lacked or nearly lacked the western harvest mouse and prairie vole but supported the thirteen-lined ground squirrel, which in turn was absent from riparian woodlands.

Extensive studies of the breeding birds of the shortgrass high plains have been carried out by J. A. Wiens, among others. In eastern Colorado's Pawnee National Grassland the eight most characteristic breeding bird species of the shortgrass plains are the mountain plover, horned lark, western meadowlark, lark bunting, Brewer's sparrow, McCown's longspur, and chestnut-collared longspur. The vesper sparrow, mourning dove, and common nighthawk are also typically present but in smaller numbers. Heavy summer grazing has been found to have major deleterious effects on breeding populations of the western meadowlark, lark bunting, Brewer's sparrow, and chestnut-collared longspur, whereas heavy winter grazing especially reduces successful breeding by mountain plover and the two longspurs. Both longspurs are fairly common nesters in parts of the Nebraska Panhandle, but the mountain plover is at the very edge of its range there and is perhaps only an accidental breeder.

The shortgrass plains of western Nebraska still support scattered and dwindling populations of black-tailed prairie dogs, although the days of vast prairie dog towns, uncounted bison herds, gray wolves, and black-footed ferrets are forever over. Brown-headed cowbirds now associate with domesticated cattle instead of bison, and lark buntings sing above alfalfa fields rather than over grama and buffalo grass. Western-bound tourists, who usually stop long enough during hot summer days to look for the still-visible wagon ruts of the Oregon Trail and to examine rusting pioneer artifacts in the air-conditioned comfort of the visitors' center at Scotts Bluff National Monument, often remain unaware of the white-throated swifts busily foraging overhead or of the occasional soaring ferruginous hawk that is as much a symbol of the High Plains as Scotts Bluff itself.

CHAPTER FOUR

Southern Shores
The Loess Hills and Platte Valley

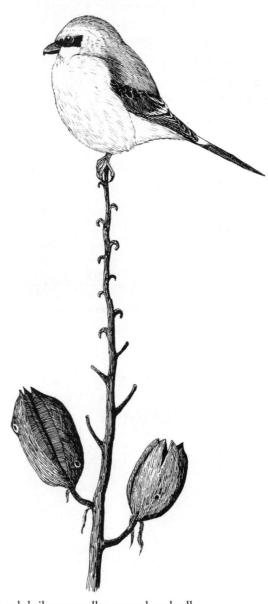

Fig. 10. Loggerhead shrike on small soapweed seedstalk.

THE SOILS OF NEBRASKA are foreign. They came originally from silty sediments of rivers draining the eastward-facing ranges of the ancestral Rocky Mountains. These sediments were later extracted from the surface layers of Tertiary and Pleistocene deposits and wind-carried up to several hundred kilometers, probably during dry and windy interglacial periods. They drifted mostly eastward and southward, to be deposited blanketlike over much of southeastern Nebraska and parts of northern Kansas and extreme western Iowa. They are now mainly to be found between the Sandhills and the lower lands lying somewhat farther east toward the Missouri Valley, where later glacial till subsequently covered or variably replaced them, obscuring their windy origins. In some places in eastern Nebraska and in the loess hills of extreme western Iowa, this silty blanket extends to a depth of as much as 50 meters. Water is readily able to penetrate such uniformly sized substrate materials, and the calcium-rich soils that have developed above them are not only very uniform in texture but also relatively low in organic matter.

Over time, wind and water erosion in the loess hills region have produced a pattern of gently undulating valleys, hills, and small bluffs. The steeper slopes of the hills and bluffs often take the form of small but nearly vertical-walled canyons, with many vertical cleavage fissures in the soft loess substrate. Along these cleavages the loess tends to break and slide downward repeatedly, eventually producing a distinctive "cat-step" erosion pattern that resembles a series of natural, nearly horizontal, terraces or steps. Overgrazing by large mammals can easily aggravate these erosive effects.

This region, traditionally described geographically as the Loess Hills and Plains (or Rolling Hills and Dissected Plains) of Nebraska, covers several thousand square kilometers of the state and is roughly bisected by the Platte River. A general geographic transition area between the shortgrass plains and sandhills to the west and the Missouri Valley to the east, it is also a botanical transition area. Although the native vegetation of the loess hills is largely mixed-grass prairie, some local areas are dominated by short grasses, and others—usually on north-facing slopes and on lowlands—approach typical tallgrass prairie. These more mesic sites, like the tallgrass prairies to the east, were among the first to be plowed and converted to dry-land farming. Many of the other areas of the Loess Hills are still only marginal for nonirrigated crop growing but provide important rangeland resources for regional farmers and ranchers.

Soon after leaving the High Plains of the Panhandle, the Platte River cuts sinuously through the heart of the loess-dominated region of southern Nebraska. In so doing, its North Platte branch brushes against the southernmost edges of the Sandhills in Garden, Keith, and Lincoln Counties. Farther along its way, it provides the eastern parts of the state with an invaluable supply of water for irrigation, industry, and domestic use, to say nothing of offering critical habitat for a few million waterfowl and cranes every spring and fall. The Platte probably represents the present-day remnant of the great paleorivers that once flowed east from the Rockies, bringing with them the raw materials that essentially formed the state of Nebraska. At least as early as late Pliocene times (about 2.5 million years ago) a river following the approximate course of the South Platte was probably feeding such materials into central Nebraska. The present channel of the North Platte, which now merges with the South Platte in present-day Lincoln County, may have been established by early Pleistocene times, or about 1.5 million years ago. The deep valleys cut across Nebraska by some pre-Pleistocene rivers were periodically dammed by glacial ice during the colder periods of the Pleistocene, causing their impounded valleys to fill with clay, sand, and gravelly materials. As a result, deep alluvial deposits now cover much or all of the Platte Valley, and these in turn lie directly above the deep sandy deposits of the Ogallala aquifer.

The present-day Platte River beyond the confluence of its two major branches is a mere shadow of its former self. In part this diminution reflects long-term climatic changes occurring before any human influence was possible, and in part it has resulted from human manipulation of its natural flows. Dams constructed on the North Platte River system during this century had, by 1960, impounded about 7 billion cubic meters of water. Most of these are in Wyoming, where there are seven reservoirs; they include the Seminole and Pathfinder Dams, each of which can impound a maximum of about 1.2 billion cubic meters of water. The most significant single manipulation of the North Platte's flow, however, is the five-kilometer Kingsley Dam and its associated reservoir, Lake McConaughy, in Keith County, Nebraska. Built during the 1930s and finished in 1941, this is the second-largest earth-filled dam in the world, and was designed to hold a maximum of about 2.2 billion cubic meters of water.

Lake McConaughy is easily the largest body of water in Nebraska, having a total surface area of more than 121 square kilometers, a shoreline distance in excess of 160 kilometers, and an overall impoundment length of about 40 kilometers. Below the Kingsley Dam there is a smaller (260-hectare) and very shallow reservoir, Lake Ogallala, which is impounded by the much smaller Keystone Diversion Dam. This concrete structure's two independently controlled spillways provide a mechanism for splitting off a significant part of the Kingsley releases (about 345 million cubic meters annually) into the first of a long series of irrigation canals. These canals result in the gravity-based irrigation of about 53,000 hectares south of the Platte in Lincoln, Dawson, Gosper, Phelps, and Kearney Counties, provided by the Central Nebraska Public Power and Irrigation District. Any remaining water may then be shunted back into the Platte's original channel. A separate Nebraska Public Power District also provides irrigation water to about 32,000 hectares in the Platte Valley. As of 1975 about a million hectares of land were being irrigated by waters of the North Platte, South Platte, and Platte Rivers. The Central Nebraska Public Power and irrigation District sold water to farmers under contract at a rate of $19.75 per 1,850 cubic meters (1.5 acre-feet) in 1991. The pricing structures vary by individual district for farms in irrigation areas controlled by Nebraska Public Power.

Lake McConaughy has flooded about 130 square kilometers of North Platte Valley habitat, leaving only a few kilometers below the dam to show clearly how this part of the valley must have looked before its impoundment. The University of Nebraska's Cedar Point Biological Station is situated on nearly 100 hectares of land hugging the southern shoreline of Lake Ogallala. Directly behind the station grounds a series of protruding bluffs and intervening steep canyons rise sharply for nearly 70 meters, their rocky north-facing slopes rather densely vegetated with eastern redcedar. The loess-based uplands that cap these slopes are dominated, when not farmed, by shortgrass (on the drier sites) or mixed-grass prairies (on more mesic sites). Between the redcedar woodlands are drier areas of mixed-grass prairie plants, sprinkled with a liberal abundance of small soapweed and prickly-pear cactus. The marshy northern shorelines of Lake Ogallala merge into low meadows of native grasses or introduced pastureland species, and within a mile or two these meadows grade into typical and relatively pristine Sandhills prairie (figure 11).

Below the diversion dam a generally well-developed riparian forest of

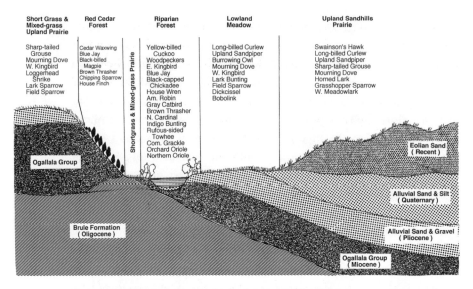

Short Grass & Mixed-grass Upland Prairie	Red Cedar Forest		Riparian Forest	Lowland Meadow	Upland Sandhills Prairie
Sharp-tailed Grouse Mourning Dove W. Kingbird Loggerhead Shrike Lark Sparrow Field Sparrow	Cedar Waxwing Blue Jay Black-billed Magpie Brown Thrasher Chipping Sparrow House Finch	Shortgrass & Mixed-grass Prairie	Yellow-billed Cuckoo Woodpeckers E. Kingbird Blue Jay Black-capped Chickadee House Wren Am. Robin Gray Catbird Brown Thrasher N. Cardinal Indigo Bunting Rufous-sided Towhee Com. Grackle Orchard Oriole Northern Oriole	Long-billed Curlew Upland Sandpiper Burrowing Owl Mourning Dove W. Kingbird Lark Bunting Field Sparrow Dickcissel Bobolink	Swainson's Hawk Long-billed Curlew Upland Sandpiper Sharp-tailed Grouse Mourning Dove Horned Lark Grasshopper Sparrow W. Meadowlark

Ogallala Group

Eolian Sand
(Recent)

Alluvial Sand & Silt
(Quaternary)

Brule Formation
(Oligocene)

Alluvial Sand & Gravel
(Pliocene)

Ogallala Group
(Miocene)

Fig. 11. Profile view of typical North Platte Valley habitats near Cedar Point Biological Station, and some associated breeding birds.

deciduous tree species occurs rather discontinuously over the sandy and gravelly bottomlands of the historically wide North Platte channel. These bottomland forests consist mostly of eastern cottonwood, American elm, green ash, various other native species, and some introduced ones as well. Adjoining the river, stands of willows have developed over sandy bars and shorelines, and where the bottomland is highly gravelly, the riparian forest is relatively open or grovelike, with cactus and other xerophytes growing in the sunnier sites between the trees.

Because all these different habitats occur within a fairly small area, the biological diversity of this location is particularly great. Of Nebraska's 60 species of reptiles and amphibians, more than 40 percent have been recorded in Keith County. The bird list for the Lake McConaughy and Lake Ogallala area consists of some 290 species, representing about 70 percent of the total avian species list for the entire state. The vascular plants of Keith County have also been well studied; they comprise some 600 species, or about a third of the total number known for Nebraska. Most of Keith County's vascular plants are herbaceous perennials, with fewer herbaceous annuals and very few woody perennials and biennial species.

Around the station grounds, major habitat types—low prairie meadows, overgrown pastures, cedar woodlands, riparian deciduous forests—attract several quite distinct avian breeding populations (table 2). In general, the relative proximity to surface water and the degree of three-dimensional vegetational structuring characteristic of the individual communities seem to have strong effects on breeding bird diversity and on overall avian abundance. Of these four habitat types, the lowland meadows has been found to have the smallest species diversity and breeding bird abundance. The deciduous floodplain forest, which exceeds the redcedar woodland in height, plant species diversity, and life-form diversity, easily claims the largest species diversity and abundance. Similarly, Craig Faanes has found that isolated woodlands in North Dakota remnant prairie areas provide important breeding microhabitats for many bird species—such as the rufous-sided towhee, house wren, and American robin—that would not otherwise breed in typical pure grassland sites. Furthermore, Faanes found that the total number of species breeding in such wooded habitats averaged about 13, or well above the number typical of North American prairie habitats generally (figure 12).

Although some cranes use the upper end of Lake McConaughy in the vicinity of Lewellen, the westernmost of the remaining major spring staging areas for sandhill cranes occurs about 50–80 kilometers below Kingsley Dam, between the towns of Sutherland and North Platte. This area has now become the least significant of the cranes' present-day spring staging areas in Nebraska's Platte Valley. Probably because current water flow rates are too scanty, the North Platte cannot achieve adequate channel width to produce the fairly vegetation-free sandbar conditions, resulting from spring ice scouring, that cranes need. In fact, ideal roosting habitat for cranes no longer exists along most of the Platte Valley. Yet despite these ecological changes, an 85-kilometer (53-mile) stretch of the central Platte River between Lexington and Denman has been identified as critical habitat for migrating cranes. So designated for the endangered whooping crane, this area also provides the most important remaining roosting habitat in the central Great Plains for the three races of sandhill crane that have spring migratory stopovers here and collectively supports the largest single seasonal concentration of cranes in the world.

Along with losses in riparian crane habitat have come losses in breeding habitat for the piping plover and least tern; the bare-sand substrates that both these threatened species need for nesting have become rare as

Table 2 Breeding birds of North Platte Valley grassland and forest habitats

Species	Lowland Meadow[a]	Overgrown Pasture[b]	Redcedar Woodlands[b]	Floodplain Forest[b]
	Density/Acre			
Western meadowlark	0.33	0.38	p	p
Mourning dove	0.16	0.38	0.56	0.66
Red-winged blackbird	0.14	—	—	—
Bobolink	0.11	—	—	—
Lark sparrow	p	0.53	0.54	0.15
Field sparrow	—	0.25	—	0.20
Northern cardinal	—	—	0.33	p
Western kingbird	—	0.23	0.28	0.10
Chipping sparrow	—	0.20	p	0.31
Blue jay	—	p	0.23	0.18
Orchard oriole	p	—	0.20	0.51
Yellow warbler	—	p	0.13	0.13
Black-billed magpie	—	0.13	p	p
Brown-headed cowbird	p	0.13	—	0.30
Northern mockingbird	—	0.10	—	—
House wren	—	—	—	0.46
Northern oriole	—	—	—	0.33
Rufous-sided towhee	—	—	—	0.30
Brown thrasher	—	—	—	0.23
Common grackle	p	—	—	0.15
Gray catbird	—	—	—	0.13
Eastern kingbird	p	—	p	0.13
Northern flicker	p	—	—	0.10
Red-tailed hawk	—	—	—	0.10
Wild turkey	—	—	—	0.10
	Totals			
Breeding species	8	9	7	18
Marginal species	12	12	18	22
Total species	20	20	25	40
Breeding density				
Birds/acre	0.83	2.46+	2.27+	4.57+
Birds/40 hectares	80	243+	224+	452+
Typical of habitat[c]				
Breeding species	2–12		8–28	9–41
Pairs/40 hectares	20–120		150–550	100–750

Note: "Marginal" species (those present at estimated densities of less than 0.1/acre) are shown as "p" if they occur at higher densities in any other habitats, but they are otherwise excluded from the table. These may include apparent visitors, late migrants, and other erratically occurring species. Organized from grassland- to floodplain forest–adapted species.

[a] Based on a 14.5 hectare (36-acre) site in Keith County (two-year average).

[b] Based on a two-hectare (4.95 acre) site in Keith County.

[c] Based on Udvardy's (1957) analysis of 21 grassland sites, 56 coniferous forest sites, and 130 temperate deciduous forest sites throughout North America.

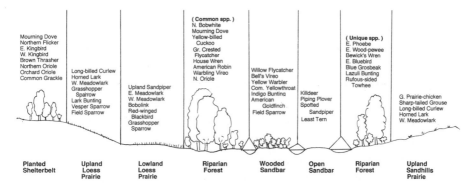

Mourning Dove
Northern Flicker
E. Kingbird
W. Kingbird
Brown Thrasher
Northern Oriole
Orchard Oriole
Common Grackle

Long-billed Curlew
Horned Lark
W. Meadowlark
Grasshopper
Sparrow
Lark Bunting
Vesper Sparrow
Field Sparrow

Upland Sandpiper
E. Meadowlark
W. Meadowlark
Bobolink
Red-winged
Blackbird
Grasshopper
Sparrow

(Common spp.)
N. Bobwhite
Mourning Dove
Yellow-billed
Cuckoo
Gr. Crested
Flycatcher
House Wren
American Robin
Warbling Vireo
N. Oriole

Willow Flycatcher
Bell's Vireo
Yellow Warbler
Com. Yellowthroat
Indigo Bunting
American
Goldfinch
Field Sparrow

Killdeer
Piping Plover
Spotted
Sandpiper
Least Tern

(Unique spp.)
E. Phoebe
E. Wood-pewee
Bewick's Wren
E. Bluebird
Blue Grosbeak
Lazuli Bunting
Rufous-sided
Towhee

G. Prairie-chicken
Sharp-tailed Grouse
Long-billed Curlew
Horned Lark
W. Meadowlark

| Planted Shelterbelt | Upland Loess Prairie | Lowland Loess Prairie | Riparian Forest | Wooded Sandbar | Open Sandbar | Riparian Forest | Upland Sandhills Prairie |

Fig. 12. Profile view of typical Platte Valley habitats, and some associated breeding birds.

vegetational succession has gradually overtaken the Platte's sandbars and sand-lined channels.

Although the species needing open sandbars and islands have suffered from increasing vegetational succession and forest encroachment along the river, however, riverine forest-adapted species have correspondingly prospered. Studies by G. Krapu and other U.S. Fish and Wildlife Service biologists between 1978 and 1980 indicated that the most abundant breeding bird species then present along the North Platte and Platte Rivers was the prairie-adapted western meadowlark, but two of the next four most common species (common grackle and mourning dove) are largely woodland-dependent. Wooded river-channel islands were additionally found to support an estimated mean density of 212 breeding pairs of birds per 40 hectares and a grand total of 35 probable breeding species. Lowland riparian forests supported a similar mean breeding density of 202 pairs per 40 hectares and a total of 50 probable breeders. House wrens, mourning doves, and American robins were the most abundant species in lowland riparian forests, and this community type also supported several unique breeders, such as eastern wood-pewee, eastern phoebe, eastern bluebird, and blue grosbeak. The cliff swallow and common yellowthroat were reported as the most abundant breeders on river-channel islands, although the cliff swallow breeds almost exclusively on bridges or large culverts in this area, using the airspace over the river and its associated islands for aerial foraging.

By comparison, sample lowland native prairie areas in the Platte Valley area supported a mean density of 47 pairs per 40 hectares and 27 probable breeders; upland native prairie habitats, a mean density of 39 pairs per 40 hectares and 31 probable breeders. Western meadowlarks, grasshopper sparrows, and dickcissels were the most common breeders in lowland and upland native prairies; eastern meadowlarks in the central Platte Valley were exclusively restricted to lowland prairie. Other species especially typical of upland native prairie were the greater prairie-chicken, long-billed curlew, horned lark, lark bunting, and—besides the grasshopper sparrow—vesper, lark, and field sparrows. The Brewer's and Cassin's sparrows were also reportedly breeding in this habitat, but the former is rare east of the Panhandle, and the latter is considered only an accidental breeder in the state.

When the Kingsley Dam was constructed, it was given a 50-year operating license, which expired in June 1987. Since then it has been operating on a year-to-year basis, pending the issue of a new license by the Federal Energy Regulatory Commission (FERC). Because of the many claims on the Platte's water by diverse and often conflicting interests, the FERC as of 1994 had made only tentative recommendations for relicensing: operating stipulations for the Kingsley Dam to be required of the Central Nebraska Public Power and Irrigation District, and corresponding requirements for relicensing the North Platte/Keystone Diversion Dam to be required of the Nebraska Public Power District. The primary interests concerned include irrigation needs, power generation, fish and wildlife values, and recreational issues.

During the 1980s Kingsley Dam was modified to provide increased hydroelectric capabilities (50-megawatt single-turbine power generation). This change required an extensive ecological modifications of Lake Ogallala in the form of dredging, higher average water levels, and increased short-term water fluctuations, which may now exceed a half-meter of surface-level variation within a 24-hour period. These activities, which remarkably enough were not formally challenged by state or federal environmental agencies, had major deleterious effects on the fish and wildlife populations of Lake Ogallala. Within a year the large populations of sora and Virginia rails that had nested in the shallow marshy areas of the lake totally disappeared, as did regular summer use and probable occasional nesting by ruddy ducks and redheads. A colony of black-crowned night-herons disappeared from a wooded island in Lake

Ogallala, and the vast numbers of marsh wrens, red-winged blackbirds, and yellow-headed blackbirds nesting on Lake Ogallala were greatly reduced. The remaining populations of these last three species now rarely manage to breed successfully, because short-term water fluctuations almost always flood their nests before the young can be fledged. The occasional pairs of endangered piping plovers that have tried to nest on the sandy islands of Lake Ogallala have likewise invariably been flooded out early in the nesting season. Additionally, increased flows of low-oxygen content waters from the deep layers of Lake McConaughy into Lake Ogallala have had disastrous effects on the trout populations of Lake Ogallala and on the sport fishing opportunities there. Midge populations, critical to the food chains of many birds and aquatic organisms, have also declined noticeably in recent years.

A few years ago Nebraska's irrigation interests convinced the state that it should challenge the right of the nonprofit Platte River Trust—founded with environmental mitigation money to protect critical whooping crane habitat—to become involved in litigation concerning Platte River water flow requirements. Fortunately, the federal government has agreed with the trust's position, stating in spring 1993 that curtailing the trust's legal rights would reduce the government's ability to enforce the Endangered Species Act. At the same time, the American Rivers Conservation Council named the Platte one of the ten most endangered rivers in North America, on the basis of its importance, the degree of threat to it, and the imminence of that threat.

The fate of the proposals affecting the rules for the continued operation of Kingsley Dam will be important not only to the locally breeding species of fish and wildlife but also to hundreds of migratory bird populations. These include the sandhill cranes, which may breed as far away as Siberia but in spring depend on the Platte for nighttime security and for easy access to a large food supply, waste corn, during a time when they must build up fat reserves for their continued spring migration to breeding grounds thousands of miles away in remote arctic tundra. It was not until 1993 that the Nebraska Game and Parks Commission took the first legal steps to assure the minimum instream flows needed on the central and lower Platte to preserve the aquatic communities of the river itself, its nearby wet meadow habitats, and critical wetland habitats for endangered or threatened species of cranes, terns, and plovers. After extended public hearings the commission's proposals may eventually reach the state's Department of Water Resources for consideration and possible action.

Decisions as to the operating requirements for Kingsley Dam in fact affect virtually all Nebraskans, because the cities of North Platte, Kearney, Grand Island, and Lincoln—or about a third of Nebraska's entire population—all depend directly on the Platte for water supplies. Beyond those concerns changes in project operation could have regional effects on locally generated electrical power. Levels of nitrates and agricultural pesticides entering Platte River waters in the highly agriculturalized areas downstream could also be directly affected. Such contaminants have important health implications in cities as far away as Grand Island (which already has significant concentrations of nitrates and other dangerous contaminants in its municipal water supplies) and possibly even Lincoln. Indeed, at least ten Nebraska towns and cities now have seriously contaminated groundwater supplies. Because surface waters and groundwaters in Nebraska are part of an interacting continuum, the quantity and quality of the water flowing through the Platte River system should be of direct and immediate concern to every resident of the state. Future decisions about the Kingsley Dam will have a direct impact on the vast majority of Nebraskans, to say nothing of their equally great effects on fish and wildlife.

CHAPTER FIVE

Eastern Shores

True Prairie and Tall Cornfields

Fig. 13. Male greater prairie-chicken.

TRUE PRAIRIES make excellent cornfields. As a result, Nebraska now produces about a tenth of the nation's total corn crop (1.07 billion bushels in 1992, generating $2.3 billion). Nebraska's cornfields also help feed the nation's second-largest population of cattle, thus promoting an interrelated water-corn-cattle economic system that has provided the foundation of the state's agriculture but has also produced serious ecological problems.

The cornfields of the present day were the native bluestem prairies of the past century, when a vast ocean of tall grasses greeted the early explorers and first homesteaders. The tallgrass prairie or so-called true prairie, bounded on the east by the riparian deciduous forests of the Missouri Valley, extended from the eastern edges of the Sandhills in Knox and Antelope Counties southward across the state to the Kansas line. Westward, the true prairies graded into the slightly drier midgrass prairies of the loess hills.

In eastern Nebraska as elsewhere throughout its broad range, true prairie assumes two general forms, that of the moister but well-aerated deeper soils of the lowlands, and that of the somewhat drier and sunnier uplands. Lowland prairies of Nebraska are invariably dominated by big bluestem, the adjoining upland prairies by little bluestem. Big bluestem forms a continuous ground cover, reaching a height often in excess of two meters at maturity, whereas little bluestem may be about a meter tall at maturity and is much more bunchlike in its growth habit. In most areas these two species co-occur in varying proportions, often reaching about equal abundance on middle hillside slopes. In the fall they turn these hillsides a glorious Indian red tinted with the coppery tones of other associated prairie grasses, especially Indiangrass.

The grasses of true prairie not only stand tall but also have deep and spreading roots. The belowground biomass of true prairies is often two to four times greater than the aboveground biomass, and the net primary production of new organic matter in prairie root systems typically represents about 40 percent of the total. Additionally, whereas the aboveground crown and shoot material of prairie herbs and forbs "turns over"—dies—every year or so, the average turnover period of prairie grass roots has been estimated at three to four years. Although most of the roots of dominant grasses of upland prairie such as little bluestem are concentrated in the upper meter or two of soil—indeed 80–90 percent of the root biomass is typically in the top 30 centimeters or so—

the taller perennial grasses have roots that often approach three meters in maximum depth. They may also spread out laterally to nearly a meter in the upper soil layers, binding together a nearly impenetrable matrix of living materials, organic matter, and inorganic materials. Such sods were strong enough to form the walls and roofs of many pioneer houses in Nebraska, some of which are still functional as much as a century later. Many tall prairie grasses also generate underground stems or rhizomes that help propagate the plants by lateral growth and serve as food reserves. Their underground location effectively protects them from prairie fires, most insect ravages, and grazing by large mammals. Above-ground nutrients produced during the summer are stored over the winter in these underground systems, permitting a very rapid resumption of spring growth in rhizome-producing grasses.

In addition to big bluestem and Indiangrass, some of the other more conspicuous rhizome-producing tallgrass species are prairie cordgrass and switchgrass. It has been roughly estimated by J. E. Weaver that an acre (0.4 hectare) of native big bluestem sod may contain as much as 400 miles (640 kilometers) of rhizomes! Similarly, a square foot (0.09 square meter) of sod may easily have 50 feet (15 meters) of big bluestem or switchgrass rhizomes, or 90 feet (27 meters) of prairie cordgrass rhizomes. The interlaced rhizomes of big bluestem and prairie cordgrass may be able to withstand about 50 pounds (23 kilograms) of stretching pressure before breaking, and those of switchgrass have been found to exhibit even greater tensile strengths of up to 132 pounds (60 kilograms). No wonder the early plows of the pioneers were unable to cut through this tough matrix!

After having been plowed and cultivated, native prairie takes a very long time to be rejuvenated through normal processes of plant succession. A study by Mary Bomberger and others compared old-field successional changes on a famous area of Nebraska tallgrass prairie, Nine-Mile Prairie near Lincoln, with similar successional patterns observed in typical Sandhills grassland (Arapahoe Prairie in Arthur County). After 40 years of succession on sites affected by major disturbance, the communities that developed in both areas were superficially similar but significantly different botanically from immediately adjacent undisturbed communities: only 12 to 18 plant species were present; and the major prairie dominant—big bluestem—was still absent from the tallgrass stand at Nine-Mile Prairie, as was a major dominant—sand reedgrass—from the Sandhills prairie stand on Arapahoe.

The dense root systems not only prevent native prairie from being easily disrupted but also make it highly resistant to invasion by other plant species. The prairie community's evolutionary history goes back at least 25 million years, to Oligocene times. During that long period successful plant species gradually evolved niches within the prairie, whereas others were no doubt outcompeted and gradually eliminated. The nongrassy perennial plants of present-day native prairie have a greater taxonomic diversity than the grasses; they comprise well over a hundred species in many prairies. Indeed, a single square mile (2.59 square kilometers) of native Nebraska prairie may sometimes have as many as 250 plant species present—approximately half the species diversity of the entire Sandhills region—and about 90 percent are perennials whose individual plants may have average lifetimes of 25 years or more. The upland prairie forbs total approximately 200 species; these have root systems similar to those of the grasses, and the most common and conspicuous prairie legumes such as leadplant are of special ecological importance because of their unique nitrogen-fixing capabilities and their consequent effects on prairie soil fertility.

Most perennial bunchgrasses of the true prairie reproduce vegetatively rather than by seeds; nevertheless, seed production can be very high, as observed by M. A. Potvin. For example, the annual seed production in a Missouri tallgrass prairie area has been estimated to be two orders of magnitude (roughly 100-fold) greater than that of nearby Missouri forest sites and other nongrass communities. This annual "seed rain" makes prairies ideal habitats for small rodents and seed-eating birds. Moreover, in contrast to the wood-dominated aboveground plant parts of forests, nearly all the annual aboveground productivity of grassland can be readily reached and grazed. Most of it can likewise be easily digested by herbivores, although—given the silica-rich epidermis of the grasses—they may need special abrasion-resistant tooth structures. Additionally, the grasses that dominate prairies have basal rather than terminal growing points (meristems), so that repeated grazing does not interfere with further plant growth. And because ground-level growing points are relatively well protected from fire, periodic prairie fires tend to stimulate rather than damage grassland communities.

Birds of the tallgrass prairie are generally similar to those of the mixed-grass prairies farther to the west, as has been summarized by P. G. Risser and others. Present in both prairie types in significant numbers are the upland sandpiper, western meadowlark, grasshopper sparrow, and vesper sparrow. Species typical of tallgrass prairie but substantially rarer or absent in the mixed-grass communities include the eastern meadowlark, bobolink, and dickcissel, but the eastern meadowlark, dickcissel, and grasshopper sparrow are actually the most characteristic breeding species consistently present in stands of tallgrass prairie. Those more typical of mixed-grass prairies and significantly rarer or absent in tallgrass prairie include the long-billed curlew, horned lark, Sprague's pipit, Savannah sparrow, Baird's sparrow, chestnut-collared longspur, and McCown's longspur. About two-thirds of the breeding birds of true prairie are migratory. The majority of the 23 most characteristic grassland-breeding species might be classified as omnivorous; the remainder are insectivorous.

Among the mammals, 138 native species occur within the five mostly grassland-dominated Great Plains states that are bounded by North Dakota and Oklahoma, and 102 species now occupy various amounts of true prairie in North America. About 18 of these prairie-adapted mammalian species are largely restricted ecologically to the North American grasslands. Additionally, 45 are primarily associated with forest-grassland ecotones, and 39 more might be considered ubiquitous. J. Knox Jones and others found the most typical mammals of tallgrass prairie to be the least shrew, eastern mole, Franklin's ground squirrel, plains pocket gopher, western harvest mouse, prairie vole, and (southwardly only) the hispid cotton rat. The hispid pocket mouse and the badger are considered more typical of mixed-grass prairies; and the black-tailed prairie dog, thirteen-lined ground squirrel, silky and plains pocket mice, swift fox, black-footed ferret, pronghorn, and bison are classified as shortgrass prairie mammals.

Species regarded by P. G. Risser and others as principally grassland-adapted mammals are the white-tailed and black-tailed jackrabbits, Richardson's and thirteen-lined ground squirrels, black-tailed prairie-dog, northern and plains pocket gophers, olive-backed and plains and hispid pocket mice, Ord's kangaroo rat, fulvous harvest mouse, hispid cotton rat, prairie vole, swift fox, and pronghorn. All of these are essentially herbivorous except for the carnivorous swift fox and the somewhat omnivorous ground squirrels; the majority are largely seed-eating rodents. Most of them are best considered grassland endemics. Exceptions include the more widespread pronghorn, the desert-affiliated black-tailed jackrabbit and Ord's kangaroo rat, and the Neotropically derived fulvous harvest mouse and hispid cotton rat. Nearly half the mammal species that have adapted to the North American prairie originally entered it either from the coniferous forests to the west and north or from the deciduous forest to the east, in roughly equal proportions. Deserts of the Great Basin or of the Southwest are the probable origins of a few species. The remaining species are too widespread in their present-day distributions to assign any possible ecological or geographic affinities.

Not far to the south of Nebraska's tallgrass prairies are those of northeastern Kansas. There, the ecology of Konza Prairie, a remnant tallgrass prairie of about 3,500 hectares (8,600 acres) located a few kilometers south of Manhattan, has been especially well studied by ecologists such as O. J. Reichman at Kansas State University. Konza Prairie lies in the Flint Hills, which support a belt of north-south oriented prairies that reaches from northern Oklahoma into southeastern Nebraska and is closely underlain by limestone. The shallow and rocky soils that make mechanized farming nearly impossible have helped preserve significant parts of the tallgrass prairie in a fairly pristine condition except that herds of cattle have replaced the bison that once grazed these beautiful grasslands. Of the plants on Konza Prairie, the most numerous are those of the aster family, with 65 species, followed by the grass family with 55 species. Collectively, over 70 percent of the vascular plants on Konza Prairie belong to families other than grasses.

Some 29 species of amphibians and reptiles have been found on Konza Prairie, including 13 snake species. Nine of these have at least occasionally been found in the grassland community. The four lizards of Konza are mostly associated with rock outcrops, and most of the amphibians are limited to the relatively few wetland sites.

There are 17 species of rodents present on Konza. Deer mice are by far the most abundant, followed by the western harvest mouse. Prairie voles are periodically common but probably undergo major cycles of increase and decline. The white-footed mouse is fairly common on the prairie but is more typical of the nearby woodlands. Woodlands also uniquely support such mammals as eastern woodrats and fox squirrels.

The 200-plus species of birds that have been observed on Konza

Prairie are extremely well studied and offer a good insight into the avian ecology of tallgrass prairie generally. Altogether, there are about 60 breeding species. According to a 1993 summary by John Zimmerman, the primarily grassland-adapted birds of Konza include 21 summer breeding residents, 9 additional summer visitors, six winter-only species, and 21 spring and fall migrants. Of the grassland breeders, 9 are "core" species that regularly breed on unburned prairie, and 6 of these also breed on annually burned prairie sites. A few avian species typical of grasslands, such as the brown-headed cowbird, breed also in one or more of the other major community types, but many breeding grassland species—the greater prairie-chicken, horned lark, sedge wren, dickcissel, eastern meadowlark, and the sparrows (lark, grasshopper, and Henslow's)—are quite habitat-specific. The species of the birds most consistently present in the grassland sites are the greater prairie-chicken, upland sandpiper, northern harrier, grasshopper sparrow, dickcissel, and eastern meadowlark. The ecologically highly tolerant and socially parasitic brown-headed cowbird is a regular albeit unappreciated grassland breeder.

There are also several strictly water-adapted species on Konza, and some that are especially associated with shrubby rock outcrop areas or shrubby seepage sites. A relatively large number, 42 species, are associated with the gallery forests that exist along stream courses. These forests consist mostly of oaks and hackberry and make up only about 5 percent of Konza's overall area but, like forests generally, support a substantially greater number and biomass of birds than do the adjoining grasslands. Forests are also the most stable of all the major communities of Konza in annual breeding bird densities. Some of the most abundant breeders in gallery forests, which collectively average about 30, are the northern cardinal, great crested flycatcher, red-headed woodpecker, blue jay, black-capped chickadee, northern bobwhite, yellow-billed cuckoo, and white-breasted nuthatch. The 33 most regularly occurring breeders include 1 herbivore (mourning dove), 1 piscivore (belted kingfisher), 6 bark-feeding omnivore-insectivores (mostly woodpeckers), 9 ground-and-shrub foraging omnivores, and 16 insectivores. The 16 insectivorous species include four separate foraging guilds: sallying foragers (flycatchers), canopy-level foragers, subcanopy-level foragers, and ground-and-shrub foragers.

The dickcissel, grasshopper sparrow, and eastern meadowlark all represent grass-dependent species whose nesting populations are not significantly different in annually burned and unburned prairie sites. Similar long-term usage patterns exist for the loggerhead shrike, common yellowthroat, and Henslow's sparrow. In annually burned grasslands the upland sandpiper is one of the most common birds during summer, where it preferentially forages in these burned-over sites, although it mostly nests in nearby unburned grasslands. The eastern kingbird nests in isolated trees, well above the influence of normal grassland fires, and forages in the open space above the trees and grasslands. Red-winged blackbirds also occur in great numbers on burned grasslands because of the local survival of protected cattail stands in wetter areas within the overall prairie habitat. In burned prairie sites the dickcissel, brown-headed cowbird, grasshopper sparrow, and eastern meadowlark make up about two-thirds of the total breeding bird population, and these same species represent more than half the birds of unburned prairie.

In the present-day absence of grazing effects on Konza Prairie, fire seems to be the primary factor affecting bird species diversity and abundance. It is used as a management technique to keep the prairie in its present successional state: that is, to avoid forest encroachment. Grassland fires tend to reduce avian species richness by reducing or eliminating standing but dead herbaceous vegetation that breeding birds such as Henslow's sparrows evidently need, as well as an array of birds (such as common yellowthroat, eastern kingbird, loggerhead shrike, and northern bobwhite) that need the presence of at least a few woody plants for escape cover, advertisement perching, or nest sites. Winter bird populations are also considerably lower, both in species richness and total abundance, on annually burned grasslands. Periodic burning has no real long-term effect on core grassland-dependent species, however, although Henslow's sparrows and common yellowthroat populations have been found to decline temporarily during the year of the fire.

The ecological effects of fire on prairie are numerous, and the individual organisms affected by it are greatly varied. Prairie fires typically are very hot but also very brief, often lasting only a few minutes at a particular site. Shrubs, trees, and other fire-sensitive species may be seriously affected; damage to perennial grasses, however, is minimal because their most critical parts and nutrient storage systems are located at or below the ground surface. Generally, fires occurring during spring tend to reduce the growth rates of cool-season species of grasses but promote those of warm-season perennial grasses and increase the flowering rates of many species of prairie plants. The number and biomass of insect herbivores such as grasshoppers also increase after a fire, thereby attracting insect-eating birds and mammals; the sudden release of nutrients into the soil following burning probably helps to account for this general growth-surge phenomenon. The nutrient content of plant foliage tends to increase during the next year or two, as do primary aboveground and belowground production and uptake rates of important nutrients such as phosphorus. Spring burning may also increase the amounts of available soil water as a result of decreased water loss through transpiration by cool-season grasses. Losses of total microbial activity and soil nitrogen levels also occur, however, and so spring fires every three or four years rather than annually may offer a good compromise, allowing the use of fire in the management of true prairies without doing serious damage to the overall prairie ecosystem.

In contrast to the Nebraska Sandhills prairie ecosystem, which is still essentially intact, the true prairies of Nebraska are nearly gone. A few scattered and protected remnants still remain, but scarcely any are more than two to three square kilometers (a square mile or so) in area. As the prairies have vanished, so too have most large grassland-adapted species such as the gray wolf, the bison, and the elk. Some true prairie species, such as the greater prairie-chicken, have simply retreated to the edges of the Sandhills, where they exist but in reduced numbers. Others, such as the Franklin's ground squirrel, have moved to the substitute prairie habitats of abandoned fields and overgrown cemeteries. Still others, including the gray wolf, are forever gone or, like the bison and black-footed ferret, are confined to captive collections or special preserves.

It has been suggested that much of the western Great Plains area historically was and still is better adapted to these animals than to humans, under present-day economic conditions, and might best be returned to a gigantic "commons" or parklike environment. To a large degree, the Sandhills already operate as a natural prairie commons. The region's native flora and fauna are still essentially intact, its original ecology is still largely apparent, and because of meager population and scattered landownership patterns it has few interfering roads and fences. Easily the largest such region of essentially coherent prairie still to be found in the Great Plains, it provides some of the best insights we can hope to find into North American grassland ecology.

PART TWO

Sandhills Scenes
Tales Told in Sand

CHAPTER SIX

Roads and Ranch Trails
Boots, Burrowing Owls, and Box Turtles

Fig. 14. Upland sandpiper landing on a booted fencepost.

MOST ROADS IN THE SANDHILLS lead nowhere. And that is one of their primary attractions. They tend to become more and more indecisive the farther one goes and finally disappear in sandy confusion, often at a fence or a rancher's gate. Thus, traveling on unfamiliar Sandhills trails is always a kind of adventure that frequently has an unknowable ending. The roads are also lonely. As a result, one can drive down the wrong side, if indeed there is a wrong side, for most Sandhills roads are only one-lane anyway. This may be desirable for a wildlife photographer, because it puts the driver as close to the left edge of the road as possible, and brings roadside birds a bit closer for photographing from the window on the driver's side. One can best exploit good photographic opportunities by cradling the camera in the lap and steering with the knees as needed. These are sometimes dangerous, not to say illegal, activities on most country roads but are well within the range of acceptable driving habits in the Sandhills.

When he was a young man in the early 1940s, H. Elliot McClure spent several years working as a biologist for the Nebraska Game and Parks Commission. He traveled many miles daily both by pickup and, as necessary, on horseback, surveying local populations of a variety of gamebirds, mammals, and other wildlife. Once, he noticed a dead mourning dove in the road some distance ahead, and mourning doves were a species of special interest to him. Rather than stop and get out to pick up the carcass, he simply slowed down, opened the driver's-side door, and leaned out to grab the bird with his left hand while holding the steering wheel with his other hand. Unfortunately, he lost his grip on the wheel and fell out of the truck! Not losing his wits, he quickly picked up the dove and then ran down the road after his truck, which was continuing down a slight slope and, remarkably enough, staying on the narrow road. About a hundred meters farther on he caught up with the still-moving truck and jumped in. Perhaps the driver of a vehicle McClure remembers following some distance behind him considered this event all part of a biologist's daily work.

McClure spent a total of three years in the Sandhills, studying the wildlife on a 1,200-hectare ranch located along the Calamus River, in Loup County. At that time the ranch had only two small fields planted in corn; the rest of the ranch was kept in natural vegetation. Some 225 hectares were natural hay meadows that helped provide winter foods for about 300 head of cattle. Twice a week he would drive the approximately

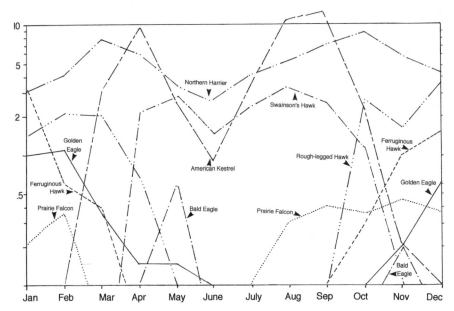

Fig. 15. Raptors (monthly mean totals) observed during 350-mile road surveys in the Sandhills. Based on tabular data in McClure 1944.

550-kilometer round trip between Ord, in Valley County, and Valentine National Wildlife Refuge in Cherry County. He also would make weekly 25-kilometer transect surveys on horseback, counting all the gamebirds and larger mammals that were visible. These road and trail surveys provide a valuable and revealing record of what the Sandhills ranch and roadside wildlife was like in the World War II years, before modernization trends began to affect both the natural habitats and the human social and economic structure of the region. Many species of birds and mammals that are now quite infrequent or rare in the Sandhills—such as golden eagles, prairie falcons, black-tailed prairie dogs, and white-tailed and black-tailed jackrabbits—were relatively common residents or seasonal visitors then (figures 15–17).

One of the first things one notices in traveling Sandhills roads nowadays is the great number of worn-out boots that have inexplicably been placed upside down over the tops of fenceposts. Some areas show this distinctive and curious landscape feature more than others. Along the

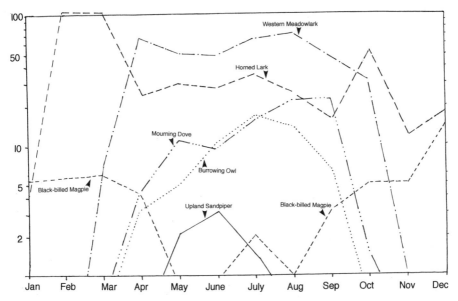

Fig. 16. Common breeding birds (monthly mean totals) observed during 15-mile trail surveys in the Sandhills. Based on tabular data in McClure 1944.

southern outskirts of Arthur, the county seat of Arthur County (where, as in McPherson and Hooker Counties, there is only a single town), no fewer than 279 boot-decorated fenceposts are visible within a distance of about two kilometers. The town itself now consists of only about 100 people, so there are nearly three times as many boots lining this stretch of fences as there are people living in Arthur. Where the boots appear less frequently, they are a minor annoyance to bird watchers, who upon seeing them at a distance keep hoping that one of these visual distractions will turn out to be a perching burrowing owl.

When the subject of decorating fences with boots is brought up, conversations with Sandhills residents and ranchers typically produce various but often unlikely or unsatisfactory explanations. Some suggest that it is simply a traditional if slightly irrational aspect of Sandhills behavior; others, that it may be a kind of territorial marking of property lines. The most logical suggestion offered is that the boots may serve to reduce the rate of decay of wooden fenceposts, which are a distinctly scarce and valuable commodity in the region—but that does not explain

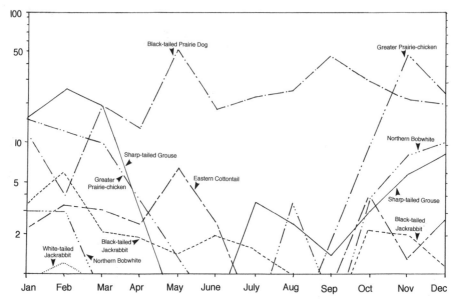

Fig. 17. Mammals and gallinaceous birds (monthly mean totals) observed during 15-mile trail surveys in the Sandhills. Based on tabular data in McClure 1944.

why they are sometimes placed over the tops of metal fenceposts. It may be that a once functional activity has, over time, become part of Sandhills culture and has lost whatever useful role it may have played.

If one wants to take a well-maintained, two-lane, north-south road through the central Sandhills, it may be well to select U.S. Highway 83. This respectable federal route connects North Platte with Valentine and is really the only choice for the traveler who is more interested in getting through the Sandhills than in actually experiencing them. Similarly, U.S. Highway 2 slants off at a somewhat tentative northwestward angle from Grand Island, headed in the general direction of Alliance. It will take one across the long axis of the Sandhills in a relatively leisurely manner, following the same general directional orientation as the drainage system of the Loup, the Sandhills' major river (figure 18).

Although generally well maintained and relatively straight, even these federal highways may run for stretches of 100 kilometers or more between gas stations or restrooms with running water. Roadside shade trees for picnicking are an even rarer attraction. But an unexpected

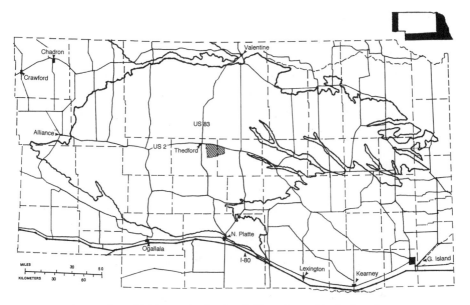

Fig. 18. Major roads and highways in the Sandhills region, with the locations of larger towns (inked) and the Bessey Division, Nebraska National Forest (hatched).

forest of trees is to be found near Thedford and the intersection of Highways 2 and 83, where the Bessey Division of the Nebraska National Forest covers almost 600 square kilometers of dunes in the heart of the Sandhills. The forest was initiated through the efforts of Charles Bessey early in this century, after the relict pine groves he found around the edges of the Sandhills convinced him that they had once been forested and could again be planted to trees.

On less well-traveled ways the presence of resting cattle in the road is a constant possibility, and going over a cattle guard at 100 kilometers per hour provides an instant indication of what a 9.0 Richter-scale earthquake might feel like if it were compressed into a fraction of a second. Sandhills roads, however, are thankfully almost free of billboards offering unwanted commercial information, unsolicited personal advice, or ominous religious prophecies. Furthermore, foolhardy speeders and drunken drivers are a definite rarity. And on all unpaved Sandhills roads at least, and many hardtop highways as well, a friendly wave of the hand to any and all approaching drivers, or at least a simple lifting of a finger or two from the wheel, is considered a minimum level of recognition for a probable kindred spirit.

Sandhills vehicles are utilitarian, to say the least; a battered pickup seems as much a part of a rancher's equipment as leather boots and a well-worn broadbrimmed hat, and a pickup's aesthetic appearance is the least important of its possibly significant qualities. In fact, Sandhill residents apparently take a perverse pride in driving vehicles that would cause some city dwellers to blush in embarrassment; one small town in western Nebraska even has an annual "ugliest truck" contest.

For biologists, driving in the Sandhills is like opening presents at Christmas. Every hill or curve offers a new vista and new possibilities. Swainson's hawks can sometimes be seen circling in the distance; northern harriers frequently course low over the valley grasses; and horned larks often scatter from the roadside ahead like leaves being swept away by a leaf blower. It is a region where every meadow has to be checked for the distinctive pockmarks of a black-tailed prairie dog town, and every fencepost has to be examined closely to see whether it is topped not by an old boot but rather by a sleeping nighthawk, a western meadowlark, or, best of all, a burrowing owl.

Burrowing owls are, like humans, only marginally adapted to the Sandhills. Their primary range lies to the west, where firmer soils and a much greater density of black-tailed prairie dogs provide optimal sources of nesting burrows. Because few prairie dog colonies of any size exist in the Sandhills proper, burrowing owl pairs are mostly scattered and confined to the occasional abandoned burrows of various other mammals, especially badger-enlarged gopher burrows. These sometimes occur along roadside ditches or, more often, in hay meadows, where soils tend to be somewhat firmer.

The best clue to the presence of a nesting pair of burrowing owls is the sighting of an adult on a lookout perch, most often a fencepost. The birds do not appear to be much bigger than the occasional western meadowlark that also often sits on fenceposts, but the owls are more upright in posture. Because of their long, spindly legs and staring yellow eyes, they rather resemble a spinster schoolmarm who is used to dealing effectively with unruly students. Burrowing owls are most active and hunt almost exclusively during daylight hours. Their hearing is not very well developed, and their nocturnal eyesight is evidently much poorer than, for example, that of the common barn-owl. Adults weigh about

150 grams, and the sexes are nearly identical in average weight, unlike most owls, among which females are usually considerably larger than males.

The burrowing owls that nest in Nebraska are somewhat migratory, but by late April the males are back on their territories. Some birds regularly return to the same nest site, or at least the same immediate area, that they used the previous year. Pair-bonding is not permanent, and either sex may occupy the nest site it used before, perhaps sharing it as frequently with a new mate as with the previous one. Probably most of these new pair-bonds are required to replace lost mates, but sometimes new mates seem to be taken even if the earlier one is still available.

Shortly after occupying a burrow, the male begins to prepare it for use by scratching around the entrance and in the interior and by lining the nest chamber with dried cow dung or other sun-dried mammalian wastes, perhaps providing a kind of olfactory camouflage. The males also begin their territorial singing, uttering a soft and frequently repeated "cuckoo" vocalization that may continue throughout the night. The male sometimes performs circular flights around his "territory" or, more probably, his home range, which may cover two or more hectares. In dense colonies, however, the nesting pairs are often situated within 10 or 20 meters of one another, with little or no indication of territorial aggression between adjacent pairs beyond the immediate nest burrow itself.

In the Sandhills, the summer foods of burrowing owls consist to a large extent of common and large insects, especially beetles. The inedible chitinous parts of dung beetles are especially prevalent in owl pellets (coughed-up, undigested food remains). These pellets also often include the skulls and larger bones of small rodents such as pocket mice, prairie voles, and kangaroo rats, and I once saw a male burrowing owl dragging an adult thirteen-lined ground squirrel back to the nest. This squirrel, which must have weighed at least as much as the owl itself, may have been run over on a gravel road about 30 meters away, but pocket gophers of comparable size have also been documented as prey. In this instance, the female quickly popped out of the burrow to collect the ground squirrel when the male arrived with his booty, then just as quickly disappeared into the burrow with it.

At the time of this particular observation in early June, at least four well-grown young could often be seen emerging from the nest entrance.

Still younger birds (the chicks hatch over a period of 10–20 days) were probably present in the nest chamber itself, but they remained invisible. After the young have grown to about one-third their fledging size, they tend to spend increasing amounts of time at the burrow entrance. There they are closely watched by the female, who sends them scurrying back down the hole at the first indication of danger.

Only a few days later a pair of badgers came loping across the field, making their way directly to the nest site. They began digging and must have been close to the nest chamber when the adult female and several young emerged from another of the burrow entrances and flew or fled on foot across the field, variously hiding in the weeds or disappearing down the holes of other old rodent burrows. The badgers probably dined on at least a few of the young owls, but certainly some survived. Badgers are probably the single most important predator of burrowing owls in the Sandhills, although bull snakes no doubt occasionally take some eggs and small young as well.

On average, about three young burrowing owls fledge from those nests that successfully fledge young, according to the observations of Martha Desmond in western Nebraska. She found that owls nesting in higher-density clusters and in greater numbers tended to be more successful than those in more scattered nests, so there should be some reproductive benefits for breeding in colonies whenever multiple nest-site opportunities make this possible. The access of a brood to other burrows—which allow the birds to spread out over a broader area for improved collective safety, or provide alternative escape routes when families are under attack by badgers—may also reduce predation losses.

At least one typical Sandhills species carries a built-in protection against most predators. This is the ornate box turtle, a distinctive, beautiful, and common reptile that may often be seen lumbering down sandy roads with utmost determination but unknowable intentions and destination. The colorful radiating yellow markings on the carapace of the box turtle make it one of Nebraska's most attractive reptiles. It is also one that can easily be captured and is all too often taken by tourists as an unusual pet. Most people think box turtles will survive indefinitely on lettuce or other vegetable materials, not knowing that they need a high-protein diet and will eventually die of malnutrition if it is denied them. Wild box turtles will eat carrion, grasshoppers, caterpillars, and probably almost anything else they are able to catch with their limited locomotor abilities, using snapping or lunging movements. Insects are probably their most important source of food, and as an important food-getting strategy box turtles seem to rely on those insects that are associated with cattle dung or on other dung-related invertebrate fauna. Home ranges of adults are surprisingly large, perhaps averaging two to three hectares, and turtles that have been displaced several kilometers from home can orient back toward it, evidently using the sun or local visual landmarks for guidance. Wandering or otherwise mobile turtles may move a hundred meters or more in a single day.

Box turtles are so named for their doubly and distinctively hinged lower plate, or plastron, which enables them to hide their vulnerable front and hind ends when danger approaches. A dog or coyote, on picking up a box turtle and finding nothing but a virtually impenetrable shell, is likely simply to drop it after a few minutes and look elsewhere for an easier meal. Young box turtles lack hinged plastrons; they can withdraw the head into the shell only by extending their hindlegs some distance out from it. Recently hatched box turtles are thus relatively vulnerable and may be eaten by crows, hawks, bull snakes and other larger predators. It is possible that raccoons have enough manual dexterity to pull open a box turtle's shell long enough to get at its softer vital parts. Badgers and possibly also coyotes can probably crack open the shell of an adult by sheer biting force. A car is likely to kill a turtle if it runs directly over it, but a glancing blow may do no permanent damage if the shell is unbroken.

Box turtles emerge from their subterranean hibernation sites in April and are probably most conspicuous in June, when adult females may be searching for suitable open nesting sites and males may be looking for mating opportunities. Males differ from females in having red rather than pale brown eyes. They are also smaller than females, and their more concave lower plastrons probably facilitate mating. Also, the first toes of males are distinctively widened, thickened, and inwardly directed, which probably assists in clasping females during copulation. Copulation occurs on land (which is unusual among turtles), and the breeding season is concentrated during the middle of the summer. Nests are dug in soft, well-drained earth, often after a week or so of searching by the female for a suitable site. Females lay at least one clutch of eggs per year, mostly in June and July; early-nesting females may produce a second clutch. Clutches average nearly five eggs, and incubation requires about

65 days. If the soil is unusually dry during autumn, the hatching of the young may be delayed until spring, or the young may escape winter freezing by burrowing into the nest's walls.

The alert eyes of a person driving down Sandhills roads may sometimes catch the quick movements of a lizard scampering across the road. The most common lizard in human-disturbed and relatively open habitats such as roadsides is the northern prairie lizard (also called the fence lizard). Widely distributed on relatively open habitats in the Sandhills, it is also often associated with blowouts or found close to soapweed clumps, under which it can quickly take cover. The lesser earless lizard, so named because of the absence of external ear openings, is also especially common in such open sites as sandy roadsides. The six-lined racerunner too likes the sunny habitats along roadsides and in disturbed areas, but it prefers somewhat denser grassy cover and tends to become inactive during the hottest part of the day. These quick animals feed on a variety of insects such as grasshoppers and beetles and, in turn, are eaten by snakes such as common racers or by small avian predators such as American kestrels and common grackles.

Sandhills roads also offer biologists some seasonal treats. Along roadside ditches a "disturbance plant community" occurs, which during summer often includes such attractive and colorful flowers as the golden yellow plains sunflower, the red-bracted sand dock or "wild begonia," the pale pink Rocky Mountain beeplant, the violet to purplish penstemons, and the white-flowered prickly poppy.

On a more somber note, one finds on all Sandhills roads the scattered carcasses of bull snakes, hog-nosed snakes, or garter snakes. Such remains, starkly contrasting with the beauty of nearby flowers, mark places where passing cars or trucks have proved to be faster than the reaction times of the snakes and should help to remind travelers that all life is transitory.

Probably the best visual treat of all in the Sandhills comes shortly before sundown. Then the shadows of the dunes play carelessly over their still-lighted slopes, creating endless yin-yang patterns to remind the viewer that light without darkness is incomplete, just as life and death are inextricably locked companions in the weft and warp of nature's rich tapestry.

CHAPTER SEVEN

Bunchgrass Prairie

Sandreed, Sicklebills, and 'Roo-rats

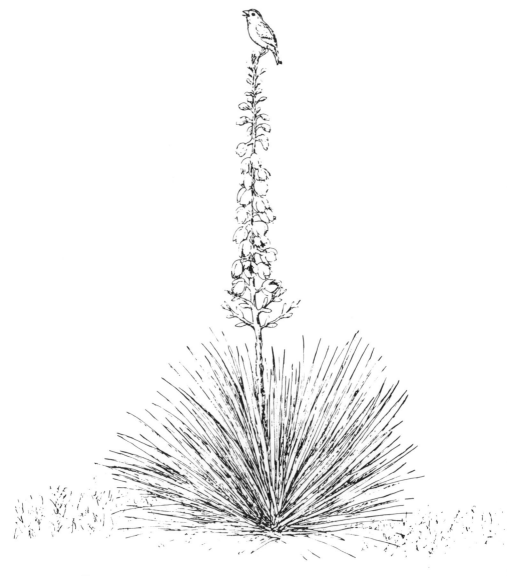

Fig. 19. Grasshopper sparrow on small soapweed.

THE KEY TO SUCCESS is resilience. The plants and animals of the Sandhills, especially those adapted to living well above the relatively moist interdune valleys, face a daily assault of drying winds and pummeling by sand grains and have had to adapt or die. The degree of adaptation that a species has to these physically demanding conditions can often be measured by its vertical distribution along dune slopes. Vegetational distributions on individual sand dunes somewhat resemble the larger-scale vegetational zones that are conspicuously stacked above and below one another along the slopes of a mountainside (figure 20). Dunes, in a way, resemble mountains in miniature, but plants living on dunes must solve some problems not faced by those that grow on steep but stable mountain slopes.

On the windward side of the dune, the constant removal and loss of surface sand particles will cause the gradual uncovering of roots near the surface and eventually threaten shallow-rooted species. Furthermore, the physical action of blowing sand is likely to do more direct damage to plant surfaces on the windward side than the leeward side, where the dune itself offers some protection. Plants with unusually flexible and small or narrow tough-surfaced leaves (such as those of most grasses) are well adapted to the windward side of the dune microenvironment.

Contrariwise, on the wind-protected leeward sides of dunes the accumulating sands that have crossed the dune crest and are falling down their slip faces will progressively tend to bury both a plant's roots and its aboveground parts. Leeward-side plants may have wider, less well-protected leaves, since they do not have to endure the constant physical abrasion of moving sand particles, but on active dunes they must be able to grow rapidly to avoid being covered by accumulating sand. They sometimes have taproots that penetrate all the way to ground water, whereas the root systems of windward-side plants sometimes tend to spread out and perhaps move or "float" with the dune.

Some plants thrive on both sides. For example, the beautifully sand-binding root system of the small soapweed may spread as much as seven to eight meters (25 feet) horizontally, and it additionally has a deep taproot that may be up to 30 centimeters thick near the surface. Even more remarkable is the tuberous enlargement of the bush morning-glory. This delicate-looking plant, with pale purplish flowers resembling cultivated morning-glories, is a wonderfully sand-adapted species: its goiterlike enlargement of the root may be more than a half-meter thick,

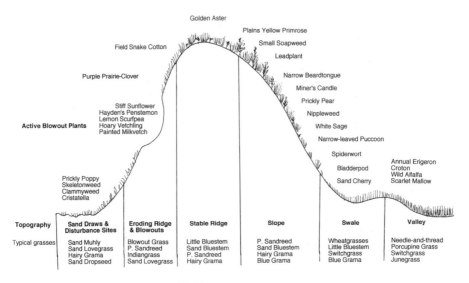

Fig. 20. Diagrammatic profile of upland sand dune communities and some representative Sandhills plants (vertical scale greatly exaggerated).

and much smaller roots reach downward at least three meters and peripherally to eight meters.

Annual plants might sometimes have a slight survival advantage on the surfaces of actively moving dunes, or around more limited blowout erosion sites, if they can complete their life cycles before being blown away. Perennials, however, are more likely to survive and compete effectively on stable dunes because of their generally deeper and more extensive root systems. Both kinds must necessarily cope with the rapid drainage and consequent high leaching rates of mineral nutrients that are associated with sand dune environments. Of the 45 typical Sandhills plants studied in 1965 by J. E. Weaver, 41 were found to have moderately deep to deep roots, and the 4 shallow-rooted species have widely spreading root systems. Laterally growing root systems are a typical feature of Sandhills plants: of the 18 most deeply rooted species, all but 4 also have widely spreading surface laterals.

In a sample of 24 grazed rangeland sites, Donald Burzlaff reported in 1962 that dry-valley Sandhills sites had a total of 91 plant species present, as compared with 82 or 83 species in gently rolling and more choppy sites. Prairie sandreed had the highest individual and collective ecological importance ranking of all plant species for the three quantitative measurements he took, and this was the case on all three site types he studied. Of the 19 species having the highest ecological importance on the basis of these measurements, 11 were perennial grasses or sedges, 3 (small soapweed, Arkansas rose, sand cherry) were shrubs, 3 (western ragweed, silky prairie-clover, prairie goldenrod) were perennial herbs, and only 2 (sixweeks fescue, annual eriogonum) were annuals.

In part on the basis of a very early analysis of Sandhills vegetation by Raymond Pool, the upland and other plant communities of the Sandhills can be classified ecologically (table 3). Pool described the plant community typical of undisturbed upland slopes, as well as of stabilized dunes in general, as a "bunchgrass association" within a larger "prairie-grass formation." This bunchgrass association is characterized by little bluestem and the less frequent but larger sand bluestem, a close (sometimes hybridizing) relative of the big bluestem of the true prairies farther to the east. Two grass species, however, are unique to the dunes: prairie sandreed and blowout grass. Both are tall-stature grasses similar to sand bluestem in height, ecological importance, and conspicuousness. All these tall grasses have widely spaced stems, well-developed rhizomes, and very deep roots.

Occurring between the tall grasses are variably smaller bunchgrasses, such as little bluestem, hairy grama, and Junegrass, as well as various forbs and smaller numbers of partially woody plants ("half-shrubs"), shrubs, and cacti. Most of the five or so typical shrubs of the dunes grow best in protected situations, and all have extensive root systems that often extend three to four meters deep. Prominent among these shrubs is small soapweed (or "yucca"), the most conspicuous single species on the upper hilltops, which thrives on the dry and windswept southern and southwestern exposures of dunes. It has a remarkable ability to resist wind erosion. Because cattle avoid eating its bayonetlike leaves, soapweed generally thrives under grazing pressure, although cattle do eat the plant's freshly grown pods. Only fire is able to reduce its vigor and abundance significantly.

In northeastern Colorado there are several areas of sand dunes similar to the Nebraska Sandhills, but the dunes are considerably lower in height (about 15 to 20 meters) and have a rather more arid climate (25–45 centimeters of annual precipitation). Francis Ramaley reported that a smaller number of vascular plant species occur there than in Nebraska's

Table 3 Native plant communities of the Sandhills region

I. Grassland (nonforested) Climax Communities
    A. Prairie-grass formation communities
        1. Bunchgrass (bluestem) association
           Dominant species: little bluestem, sand bluestem
           Principal species: lemon scurfpea, needle-and-thread, plains yellow primrose, prairie wild rose, sand cherry, sand muhly, small soapweed, shinners, stiff sunflower, sedges
           Associated topography: upland slopes, stable dunes throughout the Sandhills
        2. Needle-and-thread (speargrass) association
           Dominant species: needle-and-thread, junegrass
           Principal species: Bradbury beebalm, Platte lupine, porcupine grass, purple three-awn, white beardtongue
           Associated topography: xeric valleys and stable ridges, especially at drier edges of Sandhills
        3. Three-awn (shortgrass–Sandhills transition) association
           Dominant species: sand dropseed, three-awn grasses
           Principal species: buffalo grass, hairy and sideoats grama, needle-and-thread, pincushion cactus, porcupine grass, scurfpeas
           Associated topography: ecotone (transition) to arid shortgrass formation of high plains, western Sandhills
        4. Sand muhly (disturbance) community
           Dominant species: sand muhly
           Principal species: hairy grama, hoary vetchling, lemon scurfpea, hairy four-o'clock, painted milkvetch, prairie ragwort, sand cherry, skeletonweed, small soapweed
           Associated history and topography: sites subjected to fire or overgrazing, edges of blowouts, interdune valleys (sand draws) subject to flash floods
        5. Blowout (disturbance) community
           Dominant species: blowout grass, lemon scurfpea, prairie sandreed
           Principal species: clammyweed, cristatella, Hayden's penstemon, painted milkvetch, ricegrass, sand lovegrass, sand muhly
           Associated topography and history: sand substrate disturbance caused by wind or landslides
    B. Shortgrass formation communities: Gramma–buffalograss association
        Dominant species: buffalograss, grama grasses
        Principal species: sixweeks fescue, silver-leaf scurfpea, three-awn grasses
        Associated topography: heavy soils at northwestern edges of Sandhills having low water penetration and low annual precipitation
II. Successional Communities
    A. Weedy or ruderal (early successional) communities
        Principal grasses: wild barley, barnyard grass, annual brome grasses, purple lovegrass

Table 3 *Continued*

Principal forbs: buffalo bur, common mallow, fetid marigold, goosefoot, hemp, hoary vervain, horseweed, pigweeds, plains sunflower, ragweed, Rocky Mountain beeplant, Russian thistle, sand dock, showy partridgepea, smartweeds, sweet clover, tumbleweeds, wild lettuce, wild licorice, winged pigweed, yellow wood sorrel

Associated history or topography: old fields, roadsides, abandoned farmsteads, prairie-dog towns

B. Meadow communities

1. Hay meadow (wheatgrass) association

Associated plants: species of relatively dry meadows, especially wheatgrasses and other tall-stature, sod-forming perennial grasses

2. Rush–sedge wet meadow association

Associated plants: species of wet meadows, dominated by rushes and sedges

3. Water hemlock association

Associated plants: species of wet meadows, dominated by water hemlock

4. Fern meadow association

Associated plants: species of wet meadows, dominated by ferns

5. Willow thicket association

Associated plants: species of wet meadows, dominated by brushy willows

C. Marsh and shoreline communities

1. Smartweed association

Associated plants: terrestrial species of receding shorelines, especially smartweeds

2. Streamside marsh association

Associated plants: terrestrial species dependent upon nearby streams or springs

3. Bulrush–reedgrass association

Associated plants: emergent species nearly always growing in water or close beside it

D. Aquatic plant communities

1. Water lily association

Associated plants: floating-leaf aquatics, especially water lilies

2. Pondweed association

Associated plants: aquatic plants having submerged leaves or floating and submerged leaves, especially pondweeds

3. Stonewort–naiad association

Associated plants: submerged plants forming tufted carpets in shallow, sand-bottom lakes

Table 3 *Continued*

III. Forest-dominated (Valley and Canyon) Communities
    A. Broadleaf forest communities
        1. Basswood–redcedar–ironwood–green ash association
           Dominant species: basswood, eastern redcedar, green ash, ironwood
           Other common trees: boxelder, bur oak, black walnut, cottonwood, elms, hackberry, willows
           Associated topography: riverine valleys, especially at northern and southeastern edges of the Sandhills
        2. Paper birch association
           Dominant species: paper birch
           Other common trees: as in preceding association
           Associated topography: cool box canyons at northern edge of the Sandhills in the Niobrara Valley
    B. Coniferous forest communities: Ponderosa pine association
        Dominant species: ponderosa pine
        Other prevalent tree species: eastern redcedar
        Associated topography: Pine Ridge slopes and steep canyons in the Niobrara Valley

Source: Abbreviated and somewhat modified from Pool 1914.

Sandhills, probably owing to the drier conditions of Colorado. About a third of the Colorado sandhills flora consists of true sand-adapted plants (psammophytes); the remainder are typical of the western plains and prairies generally. Most of the more common plant species occur in both Nebraska and Colorado, however, and nearly all the community types are closely comparable. The patterns of plant succession are evidently also very similar. Ramaley's hypothetical diagram of Colorado sandhills succession is approximately reproduced in figure 21 but with most of his terms modified here to correspond to those used by Raymond Pool in Nebraska. The upper part of the diagram represents typical plant successions beginning in dry (xeric) sites such as those that develop following blowouts (xeroseres), and the lower portion indicates potential succession sequences beginning in wet (hydric) sites (hydroseres). These succession sequences theoretically converge into a stable or "climax" upland community type, dominated in both states by sand bluestem, needle-and-thread, and prairie sandreed.

It is in the loose sand or blowout community, and associated early

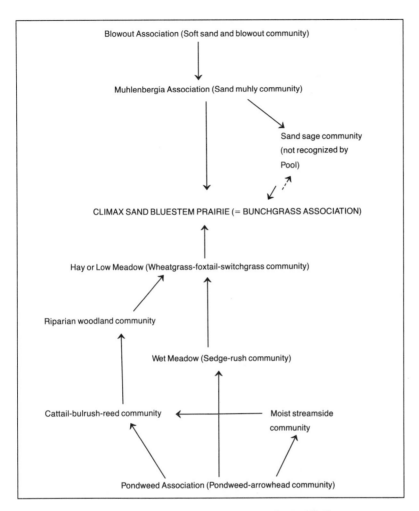

Fig. 21. Generalized patterns of plant succession in the Sandhills.

stages of plant succession, that the plants best adapted to dune living are most apparent. Five species—blowout grass, sand muhly, ricegrass, cristatella, and lemon scurfpea—are important early invaders of bare sand in both Colorado and Nebraska. One of these pioneering species, cristatella, is a weedy annual whose seeds collect in large numbers in lower areas. These germinate readily, and the plant completes its life cycle rapidly. The leguminous lemon scurfpea or "sand psoralea" also produces and disseminates large numbers of seeds that often manage,

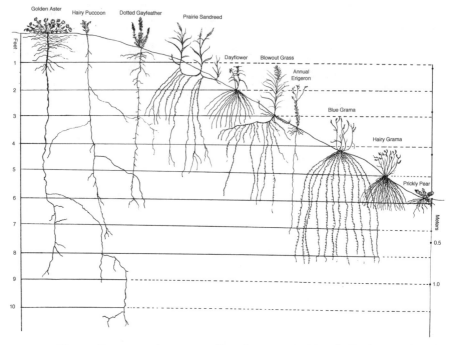

Fig. 22. Representative root profiles of some typical Sandhills plants. Adapted in part from Tolstead 1942.

like those of cristatella, to germinate and begin growth under severe conditions. Yet unlike the annual cristatella, this perennial and probably long-lived species soon establishes a vertical root system that may reach to about three meters deep, and its small roots may extend laterally to ten meters or more (figure 22). Additionally, the plant spreads laterally by rhizomes, and its ill-smelling leaves and pods are usually avoided by cattle. The other three highly sand-adapted species are all deep-rooted, rhizome-forming perennial grasses that invade loose sand primarily by lateral vegetative expansion rather than through seedling germination. They are not easily killed by being either undermined or buried by sand. They also tend to be rather coarse and wiry or to have sharp-pointed leaves that make them unattractive to cattle.

Along livestock trails, on overgrazed uplands, and in barren interdune valleys a disturbance-based community occurs, dominated by sand muhly, which is ecologically similar to that of blowouts. Sand muhly is a rhizome-producing bunchgrass that is uncommonly able to collect and hold sand, and also readily able to grow back when covered by drifting sand. Sand muhly is also a major ecological force in revegetating barren "sand draws." These are much like dry streambeds that have undergone erosion caused by both wind and water and are thus an even more inhospitable environment for plants than that associated with blowouts. Plants with well-developed rhizomes are not generally the first to invade these highly unstable and flood-prone sites, however; two annual forbs, clammy-weed and cristatella, are the earliest regular invaders. Later these may be supplemented by some of the same perennial species typically associated with blowouts. William Tolstead determined that 12 annual forbs and 2 species of annual grasses are characteristic of disturbed dune sands. Three biennial species also occur in disturbed sites, but like the annuals these are unable to compete with the xerophytic perennial grasses on undisturbed upland sites.

Around more localized disturbance sites such as blowouts, specialized and dune-adapted species—blowout grass and prairie sandreed, for example—are most likely to appear very early. Fire and overgrazing are the most common instigating causes of blowouts in the Sandhills, but the damage to plants is caused by the physical effects of moving sand, which tends to expose root systems and eventually kill the plants. Blowouts most often occur on the northwestern-facing slopes of dunes (the direction of the strongest prevailing fall and winter winds), usually somewhat below the crest of the dune. In some areas, blowouts may be as large as 45 meters in diameter and can form a dishlike depression as much as 25 meters deep as a result of downwardly deflected air currents that tend to excavate sand in a progressive, turbinelike manner. In the cratering effect thus generated, the inner blowout slope directly facing the wind is usually the longest and has a typical gradient of about 30 degrees. The opposite edge is lower, but its inner walls may be much steeper, at times almost perpendicular. The leeward slope of the dune gradually accumulates much of the sand that has been blown out of the developing crater. Various invading grasses tend to hold this loose and accumulating sand in place, often at a fairly steep slope gradient, as also occurs on the slip faces of active dunes. The capacity of a plant to recover from repeated sand burial and its ability to reproduce by rhizomes are critical components for successful pioneers in blowout sites. Sand muhly is a common invading species at these sites, but blowout grass, with its dense network of strong rhizomes, is especially effective in binding the loose sand and often begins the healing process. Sand bluestem is another common pioneer in blowouts. Eventually, prairie sandreed enters, also by means of rhizomes, and gradually becomes the single most characteristic tall-grass species to be found on the dunes. It is especially conspicuous on south-facing slopes and in the dry valleys.

In a few blowout sites the endangered Hayden's or "blowout" pen-stemon also occurs very locally. However, it is a pioneering species that cannot compete with the perennial grasses indefinitely, and it lacks the ability to spread by rhizomes or stolons. Of 34 common forbs of the bunchgrass uplands, 28 species can reproduce by rhizomes. Those forb species lacking rhizomes are typically scattered over the dunes as isolated plants and possess deep taproots, according to William Tolstead. He also reported that among 79 forb species occurring on the sand dunes, 49 are perennials, 19 are annuals, 6 are partially woody half-shrubs, and 3 each are biennials and cacti (stem succulents). He considered the five most important forb species on dunes to be the perennials spiderwort, skeletonweed, prairie sage, western ragweed, and lemon scurfpea.

At the drier western edges of the Sandhills, upland sites may be dominated by two perennial grasses: needle-and-thread and junegrass. On shallower, firmer, and less sandy soils, sand dropseed and three-awn grasses often dominate instead and provide a community type transi-tional to the shortgrass prairies of the High Plains. Small soapweed is found in these western Panhandle areas too but is most common on rough and stony lands and on heavily grazed sites.

Small soapweed survives very well under disturbance conditions and resists further erosion because of its remarkable sand-holding and sand-collecting abilities. Throughout the upland communities of the Sand-hills where small soapweed, or yucca, is ubiquitous, its cascade of ivory-white blossoms in June provides what may well be the greatest annual floral display to be seen in all of Nebraska. The large and permanently open flowers are distinctively scented, the scent being strongest at night, but only a small and inconspicuous amount of nectar is produced. These flowers hold an additional secret that is all too easily overlooked. I had first read as an undergraduate about the classic mutualistic tie between the uniquely American yuccas and a small moth pollinator: in the Great Plains, all the yuccas are pollinated by a single moth species. Wanting to see the yucca moth itself, I watched the soapweed plants

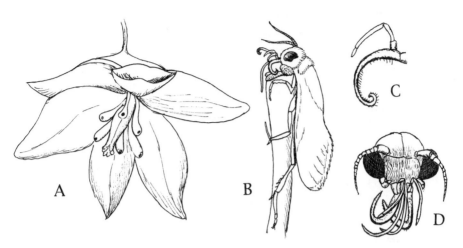

Fig. 23. Yucca flower (A), yucca moth (B), moth's maxillary palp with associated tentacle (C), and frontal head view (D).

near Cedar Point intently whenever they were in bloom but never saw a sign of any insects around them. Year after year I watched and waited, but always in vain. Then, a few years ago, after concentrating on a blooming soapweed for about half an hour near sunset without seeing any activity, I shook the plant's stem in annoyance. Immediately a few tiny white insects fell out of the blooms, so small and inconspicuous that at first I couldn't believe they were actually moths. Yet here at last was the famous yucca moth, which spends the day hiding almost invisibly within the flower and probably does whatever flying it may undertake at night.

The yuccas of the American West are entirely dependent on these tiny moths for pollination, and the moths in turn depend on yuccas for their own long-term survival. The yucca's nectar is not an attraction to or drunk by the moth; instead it may simply serve to distract nonpollinating insects from approaching the flower's stigma. When a female yucca moth lands on a flower, she first climbs one of its six long and sturdy stamens to reach the pollen on the anther at its tip. Then she prepares a large ball of pollen; indeed, her mouthparts and front legs have been specifically modified to facilitate this behavior (figure 23). The tongue is extended to steady the head, the maxillary palps are used as scrapers, and the ball is held in place between the forelegs with specialized maxillary tentacles. These tentacles, which are lacking in males and also in the closely related "bogus" yucca moths, grow out near the bases of the maxillary palps.

Pollen may be gathered from as many as four stamens of a single flower. The moth then typically flies to a separate flower and investigates the condition of its ovary, apparently to determine whether it has already been pollinated. If not, she places her pollen ball into the hollow stigma with one of her maxillary tentacles, ensuring that particular flower's pollination. She then deposits her eggs in the lower part of the same pistil, after first boring into it with her ovipositor. Typically she lays one egg each in the three cells of the flower's ovary, usually repollinating the flower repeatedly between successive egg-laying acts.

When the larvae hatch, they begin to eat the nutritious and conveniently close developing seeds. Enough are left uneaten from the potential crop of 300 to 400, however, to ensure a supply of mature and viable seeds from that flower. The fully grown larvae eventually bore holes in the side of the developing pod and complete their pupation and emergence outside the plant. The emergence of adults from a single season's brood may be spread out over the next several years, ensuring that the moth will continue to survive, even if the yucca's flowering and seed production should completely fail during some years because of late freezes or other causes. In this way, both species achieve successful long-term reproduction, and a state of obligatory mutualism or "symbiosis" has been attained. By comparison, the related bogus yucca moths cannot pollinate yuccas because they lack maxillary tentacles, and their larvae feed in the stems or fruit without any evident benefit to the plants. Presumably, the symbiotic relationship between the yucca and the true yucca moths gradually evolved from a similar, less mutually advantageous ancestral condition.

For terrestrial animals, survival on an unstable dune substrate involves some serious problems. Few berries, fruits, or succulent leaves are available for herbivorous mammals. Although grazers such as bison are beautifully adapted to grazing the abundant grasses and succulent forbs of prairies, these heavy-bodied ungulates have small, pointed hooves that tend to sink deeply into soft sand with every step, making locomotion difficult. Smaller-bodied seed-eating mammals, however, especially small heteromyid rodents such as kangaroo rats (Ord's and Merriam's) and pocket mice (hispid, olive-backed, plains, and silky) are correspondingly favored. Their physiological adaptations to desert life include the storage and consumption of relatively small, dry seeds. They have a reduced physiological need for free-water sources but instead can manufacture their necessary water metabolically; additionally, they can concentrate their urea, and their nocturnality minimizes surface water loss by evaporation. Anatomically, their elongated hind limbs and large rear feet are highly efficient on sand, as are their special locomotory techniques such as bipedal hopping (saltation) and the use of erratic escape routes (ricocheting). Flicking or kicking sand at enemies is also an effective defensive device. Sandbathing is common among dune-adapted rodents and some other desert-adapted mammals. Such behavior may possibly help to avoid excessive oil buildup in their pelage and, at least for rodents, provide scent-marking sites.

Sparse vegetative cover on sand dunes neither offers terrestrial animals a source of cooling shade nor provides protection from diurnal and visually hunting predators such as hawks. Daytime and nocturnal temperature variations are often fairly extreme in the dune environment, partly because the typically low humidity levels facilitate rapid short-term temperature changes. For the warm-blooded (endothermic) mammals, nocturnality is a common adaptation for dealing with temperature-related problems; however, the "cold-blooded" (ectothermic) reptiles that depend on external temperatures to regulate their internal metabolic rates may be unable to exploit this particular strategy and must often remain inactive, out of direct sunlight, during the hottest parts of the day. Sand is an especially difficult habitat for all burrowing animals as well: although digging in sand may be easy, the finished burrows tend to collapse equally readily.

Birds are physiologically much like reptiles in that they are preadapted for living in dry environments. They lose little water to evaporation, since they lack sweat glands; they convert their nitrogenous wastes to water-insoluble uric acid rather than urea and thus do not need to waste any water in producing urine. Their thick-shelled eggs are well adapted for incubation in dry environments—indeed, generally better adapted than those of lizards, turtles, and other egg-laying reptiles. Unlike reptiles and most mammals, they can normally escape most predators by simply taking flight and so do not need special mechanisms for getting around on sand or digging underneath it. (Few dune-breeding birds are burrowers; burrowing owls, for example, are found only on the somewhat heavier valley-bottom soils.)

Most Sandhills-breeding birds are inconspicuously colored; the grouse, sparrows, shorebirds, and horned larks are all examples. Buffs and browns are desirable colors for survival on the exposed dunes. Bright colors are best left to other, variously protected species, or effectively hidden and only briefly exposed as during sexual display, when contrasting wing or tail patterns may be exhibited or distinctive crests or pinnae brought into view.

Upland Sandhills birds are generally small, few, and far between (table 4). One limitation for birds breeding in Sandhills habitats is that there is little height to be found in the common vegetation. The only fairly tall and sturdy plants commonly present in the upland sites are old flowering stalks of small soapweed. Their stiff stems often serve as territorial advertising posts or as convenient lookouts for various small birds. Several dune-breeding species—such as upland sandpipers, long-billed curlews, and horned larks—are fairly independent of such perching sites, instead usually proclaiming their territorial occupation by means of aerial displays and associated vocalizations. Song flights are common among grassland-adapted birds generally, but poor fliers such as sharp-tailed grouse and greater prairie-chickens tend to use elevated dune tops or ridges for their collective display sites. These provide excellent visibility for approaching females but make it difficult for predators such as coyotes to approach undetected.

One famous and beautiful bird that is especially characteristic of the upland prairies of the Sandhills is the long-billed curlew. It is the second largest shorebird in the world (one Asian species of curlew is slightly larger), with a wingspread of 60 centimeters and a gracefully down-curved bill that measures as much as 20 centimeters in females and 15 centimeters in the smaller males. The scientific name *Numenius,* which translates as "new moon," refers to the shape of the bill; the common vernacular name "sicklebill" is equally apt.

Long-billed curlews have greatly suffered in recent decades from range retraction and diminished numbers almost throughout their entire breeding range; in many states of the upper Midwest where they once were fairly common they are now considered as extirpated. But in the Nebraska Sandhills they are still sufficiently common that one might

Table 4 Breeding birds of Nebraska Sandhills and tallgrass prairie habitats

| | Nebraska Sandhills | | | Eastern Nebraska Tallgrass Prairie Pairs[c] |
| | Upland Prairie | | Wet Meadows | |
	Birds[a]	Nests[b]	Nests[b]	
	Density/40 Hectares			
Ducks (*Anas* spp.)	—	—	12.4	—
Ring-necked pheasant	—	—	—	7.0
Greater prairie-chicken	—	—	0.9	—
Sharp-tailed grouse	—	1.0	—	—
Upland sandpiper	2.45	1.0	4.4	—
Long-billed curlew	1.88	—	—	—
Mourning dove	25.8	1.0	—	1.4
Horned lark	—	1.0	1.8	—
Dickcissel	—	—	0.9	5.6
Field sparrow	—	1.0	—	2.8
Vesper sparrow	—	1.0	—	—
Lark sparrow	7.8	2.1	—	—
Grasshopper sparrow	14.3	9.4	0.9	21.0
Red-winged blackbird	—	—	2.7	—
Eastern meadowlark	—	—	0.9	4.2
Western meadowlark	32.9	3.1	8.9	1.4
Brown-headed cowbird	—	?	?	2.8
	Totals			
Total breeding species	6	9	13	8
Density/40 hectares				
Nests	—	20.6	34.8	—
Birds	85.1	—	—	—
Pairs (territories)	—	—	—	47

[a]Based on a 14.5-hectare (35.8-acre) study area in Keith County near Cedar Point Biological Station (two-year average).

[b]Based on 38.9 hectares (96 acres) of choppy upland sandhills and 45.7 hectares (113 acres) of wet meadows (Schwilling 1962). Nesting ducks included mallard, northern pintail, blue-winged teal, and possibly others. Since brown-headed cowbirds are brood parasites, their nesting density cannot be judged from nest counts.

[c]Based on 28.3 hectares (70 acres) of restored prairie (Allwine Prairie, Douglas County) in eastern Nebraska (*American Birds* 37 [1983]: 83).

Fig. 24. Female long-billed curlew, calling in flight.

count 50 or more in a single hay meadow during July, when the birds are gathering prior to their fall migration. During late May and early June it is rare to spend a day in the Sandhills without seeing and hearing curlews. At that time of year the incubating or brooding birds, extremely alert to any possible disturbance, intensively mob any person or other possible threat to their nest or brood with diving calls and raucous screams from nearby dune tops (figures 24, 25).

Long-billed curlews are especially evident shortly after they arrive in April, when unpaired males perform advertisement flights above their territories. In these so-called "bounding-sκκ" (soft "kerr, kerr") flights the males rise almost perpendicularly, then set their wings in umbrella style and glide down, uttering a series of rapid and melodious "kerr"

Fig. 25. Long-billed curlew behavior: defensive flight (A), bounding soft-kerr-kerr (SKK) flight (B), aggressive-upright posture (C), aggressive crouch-run (D), precopulatory wing-raising by male (E), nest-scraping display (F), and curlew chick crouching (G).

notes. They may almost reach the ground before rising to start another calling and gliding sequence. A quite different pattern called the "arc flight" is performed when human or other mammalian intruders come near a nest or brood. The threatening bird flies directly toward the intruder, only to veer away at the last moment and begin a high arcing flight that precedes the next approach. A wing-lifting display, exposing the beautiful cinnamon-colored underwing surface, is used by adults during intense threat display, and a similar upright wing fluttering is performed by males just before copulation. Nest-scraping behavior is used during courtship and may result in the choosing of an actual nest site.

Females hide their nests in low cover on the upland Sandhills prairies, often in a clump of grass less than 25 centimeters high, or barely enough to cover the incubating bird. Yet they are virtually invisible under those conditions; I have been within three meters of an incubating female and aware almost exactly of where the nest was located, yet unable to see the bird before it flushed. Often the nest is situated near a shrub, rock, or mound of dirt, which perhaps help the female to locate her own nest. Curlew clutches are almost invariably of four eggs, and incubation takes four weeks. In the Sandhills, hatching often occurs during the last week in May and the first week of June. All the chicks emerge from their eggs within a period of five hours or so. If they hatch late in the day, they are likely to spend the night in the nest, leaving it early the following morning. The highly mobile chicks are guarded by both adults. After the young are two to three weeks old, the female often abandons her brood, but the male continues to look after them until they fledge at about six weeks of age. Fledging usually occurs before the end of July in the Sandhills, and the birds then undergo an early migration out of the region, probably heading toward wintering grounds on the Pacific coast.

If the long-billed curlew is king of the dunes among breeding birds of the bunchgrass prairies, Ord's kangaroo rat is probably the most characteristic small mammal of these sandy grasslands. Indeed, it exhibits more special adaptations for sand dune life than any of the other mammals of the Sandhills. Ord's kangaroo rats, affectionately called " 'roo-rats" by biology students at Cedar Point Biological Station, are among the most visually appealing of all Nebraskan mammals. Their large heads, enormous limpid eyes, kangaroolike body shape, and occasionally upright posture all confer an air of "cuteness" that is impossible to resist. Except for the fact that they are strongly nocturnal, they might well make the perfect pet, because they can survive without water and live on a simple diet of seeds, and their cages do not acquire the urine-drenched odor typical of most rats and mice. Kangaroo rats kept as pets often survive several years, but those in the wild are probably in general not nearly so lucky, as a variety of predators—such as bull snakes, common barn-owls, and great horned owls—help to keep their populations in check.

Kangaroo rats are members of the same rodent family (Heteromyidae) as pocket mice, and in the Sandhills the kangaroo rat and two species of pocket mice probably make up more than three-fourths of the small-mammal population of the upland grasslands. In a study by Walter Baumann of the rodents of Arapahoe Prairie, among some 770 animals that were live-trapped and individually marked the plains pocket mouse represented 68 percent, the hispid pocket mouse 3 percent, and Ord's kangaroo rat 2 percent. The northern grasshopper mouse, constituting 12 percent, made up much of the remaining sample. The population densities of the two pocket mouse species were judged to be highest in old-field (previously cultivated) grassland sites but were almost as high on upland bunchgrass slopes and lowest in valley sites. Too few kangaroo rats were captured to establish possible community preferences for that species. In studies of the food habits of rodents in eastern Colorado shortgrass prairies, L. D. Flake found that kangaroo rats there consumed only a small amount of animal materials in their diet: 4 percent by volume, compared with 74 percent of the foods eaten by the northern grasshopper mouse. Unidentified seeds made up 74 percent of the diet of kangaroo rats but only 7 percent of the foods consumed by northern grasshopper mice, showing a nearly inverse relationship in the importance of these two food groups between the two species.

As compared with the much more widespread and anatomically generalized mouse genus *Peromyscus,* pocket mice and kangaroo rats have lower rear limbs of progressively increased length, which facilitates effective bipedal locomotion (figure 26). Relative to these other two rodent genera, kangaroo rats have enormously enlarged middle ear cavities, which were shown by Douglas Webster to increase the rat's auditory sensitivity to certain weak and low-frequency (1,200–2,600 cps) sounds such as the noises made by various attacking predators: the very weak flight sounds of approaching owls, and the similarly ultrasoft mechanical sounds produced by the locomotor movements of striking rattlesnakes.

When suddenly alarmed—threatened by a snake, for example—a kangaroo rat can quickly leap upward and backward. Such quick jumps often move the animal back about 30 centimeters and about the same distance above the ground. A captive Merriam's kangaroo rat observed by G. A. Bartholomew and H. H. Caswell was able to leap about 1.5 meters horizontally and up to a half-meter high in a single bound from a

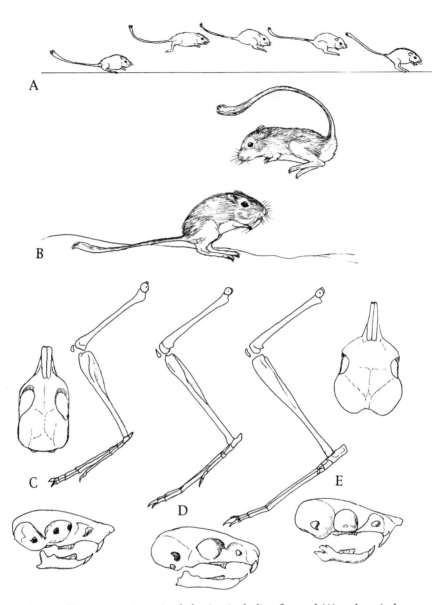

Fig. 26. Kangaroo rat jumping behavior, including forward (A) and vertical
(B) leaps. Skulls and hindleg bones of *Peromyscus* mice (C), *Perognathus* pocket
mice (D) and *Dipodomys* kangaroo rats (E) are also shown.

resting position. When jumping vertically, the animals typically hold the tail upward above the back, apparently to help maintain balance. When fleeing from ground-based predators, they perform a series of bounding movements that may carry them forward a meter or more at each leap; additionally, they make unpredictable changes in direction, including sudden turns of at least 90 degrees. During forward hopping, the slightly upcurved tail is held almost directly backward in a kangaroolike manner, and again it probably serves to maintain bodily balance.

When investigating or reaching for objects above its head, a kangaroo rat will often stand nearly upright on its hind legs. A similar bipedal posture, described as a "rigid upright" by John Eisenberg, is used by kangaroo rats for sparring. This vertical orientation grades into more horizontal configurations, including an elongate posture associated with quadrupedal locomotion and a more rounded-body posture used during resting. The resting posture is similar to the closed-eyes submissive posture. A very stretched-out posture is also associated with sand bathing, which is commonly performed by kangaroo rats and other heteromyids and, as noted, probably helps keep the pelage in good condition. The animal begins by digging into the substrate with its forepaws, then stretches forward on its belly or side while extending its forepaws as far forward as possible (figure 27). This behavior probably also serves as a scent-marking device by influencing animals of one sex to sand-bathe where those of the other sex have already done so, and in this way may provide an important social function. Thus, a male can leave his scent at a female's sand-bathing site and may become sexually excited if her odor indicates that she is sexually receptive.

When courting, two individual kangaroo rats perform various nose-to-nose, nose-to-anus, and mutual nose-to-anus contact behaviors. Copulation is achieved when the male, after following, driving, rump-patting, and grooming the female, finally grips the back of her neck and mounts her. At this time the female raises her rump and awaits completion of the act. Copulation perhaps typically occurs inside the female's burrow rather than on the ground surface. Pair-bonding beyond the moment is probably lacking. Generally, adult kangaroo rats seem to avoid one another, and closely confined adults are likely to fight to the death; it is thus unlikely that any extended pair-bonding occurs in the wild.

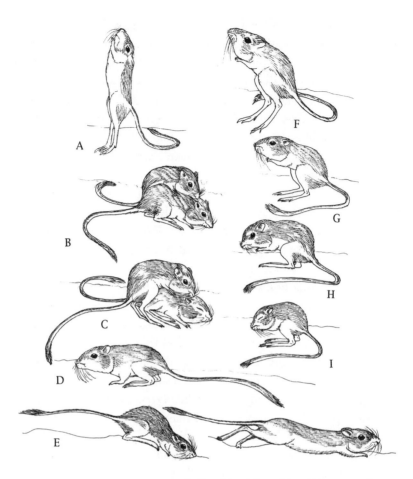

Fig. 27. Kangaroo rat postures, including bipedal reaching (A), precopulatory (B, C), elongate (D), sandbathing (E), rigid upright (F), investigatory upright (G), resting (H), and submissive (I) postures.

The kangaroo rat that consumes the seeds of the sand dune plants and the long-billed curlew that eats the grasshoppers that have eaten the grass are part of an interconnected web of energy and of species whose ecological relationships are sometimes obscure but always fascinating. These relationships are most often apparent along steep environmental gradients, which the Sandhills provide in great number and variety. The vertical distance between the upland dune community and the more mesic community occurring on the lower slopes and in the valleys between the dunes is often not very great, perhaps 30 meters or less, yet many of the vascular plants tend to sort themselves out along the ecological gradients of available moisture, drainage, and other physical and biotic characteristics that connect the dune crests with the lowland meadows. The resident small mammals and breeding birds are too mobile to be easily plotted along such minor environmental gradients, but they too probably adjust to slight shifts in vegetation types, densities, and heights and perhaps to other biological or physical characteristics of the environment. In such ways biotic communities are structured, and discerning the adaptive processes involved is a major concern of community ecologists. Rather than searching to find sermons in stone, the Sandhills ecologist may instead hope to discern evolved ecological strategies in sand.

CHAPTER EIGHT

Lowland Meadows

Grasshoppers, Grasshopper Mice,
and Grasshopper Sparrows

Fig. 28. Grasshopper sparrow on plains sunflower seedstalk.

THE PRICE OF EXISTENCE is dying. That grim realization, perhaps more than anything else, sets humans inexorably apart from all other animals and has generated a bewildering variety of often contradictory religions. Other animals seem to be able to live enthusiastically, reproduce exuberantly, and even die with a kind of ease and grace that we are wholly unable to understand. Life in nature is usually short and its ending typically abrupt; even a lucky individual's life span may be measured in hours or days rather than in years. So it is that the commonest vertebrate animals of the Sandhill's lowland meadows, the various mice and native breeding sparrows, have annual mortality rates that from a human perspective are mind-boggling. And for common grassland insects such as grasshoppers, the mortality rates of eggs, nymphs, and adults are simply astronomical. Nature's prodigal gift of abundant life must be roughly balanced by an equally severe sentence of early death. Such is the price of living and, combined with genetic diversity and the related elimination of the less fit by natural selection, a primary key to rapid evolutionary adaptation.

Still, there are some exceptions to this compressed time scale in nature. Probably the oldest individuals in the Sandhills are to be found not among the scattered old-time ranchers but among the soapweeds, the prickly-pear cacti, or perhaps even the clumps of bluestem or other perennials that seem to exist in an unwavering show of apparent fortitude and stoicism from year to year and from decade to decade. S. G. Weller's studies in Michigan dunes of hairy puccoon, a common Sandhills herb, indicate that longevity in that inconspicuous and seemingly defenseless species may sometimes approach a century.

Life seems to be somewhat more accelerated in the mesic valleys of the Sandhills than it is on the dune crests. Raymond Pool described the valley floors between the wet meadows that adjoin wetlands and the bunchgrass-dominated uplands as the "hay meadow association," because the present-day use of these areas (if they have not been converted to irrigated cornfields) is mostly as native hay meadows that are annually mowed for cattle food. In contrast to the relatively open uplands, the hay meadows constitute a closed vegetational community; that is, they consist of a continuous vegetational ground cover, with plants that usually range in height at maturity from about 60 to 120 centimeters. Many relatively showy, generally mesophytic forbs are to be found here, including composites such as coneflowers, sunflowers, and goldenrods;

a variety of mints, legumes, and equally attractive flowering plants such as prairie larkspurs and Carolina anemones are also common. The more shallow-rooted Sandhills forbs are either confined to such meadow sites or have evolved drought-evading adaptations (by completing their flowering cycles early in the summer) or drought-enduring capabilities (as have the succulent but unpalatable species of cacti).

Dominating the hay meadows are various perennial grass species such as wheatgrasses, big bluestem, Canada wild rye, switchgrass, and other relatively tall and sod-forming grasses that are common to and important components of the true prairies farther to the east. Yet despite the hay meadow's apparent richness, the botanical species diversity of valley communities at Arapaho Prairie is actually lower than that of the much more arid dune ridges. Reasons for this paradox are still uncertain but may relate to soil-controlled variations in available substrate moisture over space and time and to interspecies differences in the ability to partition this valuable resource.

From the dune crests to the swales and interdune valleys, soil and moisture relationships gradually change, with the soils becoming more fine-textured, more stable, and able to hold increasingly more water. Thus, some shallow-rooted and "cool-season" grasses such as needle-and-thread, blue grama, and porcupine grass can readily compete and survive in these locations, and these species typically dominate the more mesic valley communities. Most of the grasses growing in the relatively moist hay meadow community also occur in somewhat higher swales, but increasingly, the warm-season plants of the bunchgrass uplands appear and coexist, and a continuous biological gradient is attained. The cool-season grasses (the so-called C-3 grasses, in reference to their photosynthetic pathways for fixing carbon) have a different metabolic pathway for photosynthetic production than do warm-season (C-4 grasses), and C-3 grasses tend to exhibit higher rates of water transpiration than do C-4 grasses when there is an abundance of water in the root zone. Additionally, C-4 grasses tend to be more water-conserving and to have deeper root systems, making them better adapted to surviving on the upper dune slopes and ridges. In intermediate swale communities the cool-season western wheatgrass typically dominates the depressions wherever some organic material accumulates in the soil. This swale community then merges with the lower dune slopes, where the more xeric- and sand-adapted prairie sandreed and blue and hairy grama become increasingly important. Likewise, the hay meadow community gradually merges toward the wetter extreme, progressively grading into the subirrigated wet meadow, fen, and marsh communities that immediately surround deeper permanent wetlands.

The highly productive hay meadows are often grazed heavily by cattle. They then become exposed to invasion by more weedy grass species. And although the blowouts typical of higher dune slopes are infrequent, when overgrazed meadows are exposed to strong winds, a fairly extensive "blow-plain" may result, denuding fairly large areas and setting off a plant successional pattern similar to that associated with the typical blowouts of higher elevations. Western wheatgrass has strong rhizomes and, when sufficient moisture is present, is able to invade more loamy lowland areas that have been affected by plowing and similar disturbances; however, it cannot compete indefinitely with blue grama and needlegrass at higher and drier sites and in the absence of such disturbance.

The relative abundance of plant food provided by the lowland meadows is exploited by many kinds of herbivorous animals, not the least of which are prairie voles and grasshoppers. Prairie voles reach their highest abundance in the denser grassy sites of Arapaho Prairie, leaving the bunchgrass and forb zone to the two pocket mice and most open sandy areas to the Ord's kangaroo rat. Representatives of the four subfamilies of grasshoppers also present at Arapaho exhibit a great diversity in foraging behaviors and life history characteristics (figure 29). A total of 38 grasshopper species were observed by Anthony Joern on this fairly small area as of 1977, and 27 additional ones have been found since then. Most of these occur as adults during autumn and die before winter, but some species overwinter as nymphs and become adults during spring and early summer. The maximum number of species occurring as adults were tallied by Joern during August: 24–28 species were then present, as compared with only 4–12 in June and 20–23 in September. Grasshopper densities were highest in disturbed areas, such as around a stock tank, and next highest in valley sites; slope and ridge sites had the lowest average densities. All the most common grasshoppers were found to consume mainly grasses and sedges. Over a four-year study period the most common species of all was the white-whiskered grasshopper, which

Fig. 29. Male Haldeman's grasshopper (left) courting a female on a prairie sandreed leaf (above right) as a mermira grasshopper (below right) watches.

occurred in all Arapaho habitats but was relatively most abundant on ridge locations. Adults of this species were present from late June to at least the end of September.

One species of robber fly (figure 30) was found to be an important predator of grasshoppers, estimated by Joern to take nearly one-fourth of the adult grasshoppers present at one valley site over a 30-day period. Grasshoppers also happen to be one of the favorite foods of northern grasshopper mice. In the shortgrass prairies of eastern Colorado they were found by Lester Flake to be second only to adult beetles as primary foods, and made up about a quarter of the diet in relative volumetric terms. Grasshopper mice might be expected to be more numerous in the more grasshopper-rich valley communities than on the slopes, and a study by Marvin Maxwell and Larry Brown in eastern Wyoming indicated that they were strongly associated with grassy communities on loam and loamy sand soils, especially those dominated by blue grama and needlegrass. Walter Baumann's study at Arapaho Prairie, however, indicated that they were most common in microhabitats having relatively open vegetational spacing—that is, on slopes and on old-field successional sites—and not distributed in a way that directly corresponded with the densities of such preferred foods as grasshoppers.

Grasshopper mice prefer to locate their burrows in open sites, and their large home ranges no doubt readily allow them to reach rich foraging locations. Baumann estimated the mean size of their home ranges as about 1.3 hectares in old-field sites and 0.7 hectare on slope sites. These seem rather small figures for this widely ranging mouse; their ranges have been estimated by others such as D. G. Ruffer to average about 2.4 hectares.

Because of their highly carnivorous diet, the social behavior of northern grasshopper mice is of special interest. Unlike most rodents, they pair-bond. This trait is adaptive in predatory mammals in that it increases the probability that the female will survive to rear a brood and improves the male's chances of transmitting his own genes to the next generation—even though it also reduces his opportunities for additional matings with other females. The nest burrows are evidently excavated by mated pairs, and the boundaries of territories are scent-marked.

Like other mice, grasshopper mice become sexually mature when still very young; wild females of the southern species have been found breeding when only seven weeks old, about a week after becoming sexually

Fig. 30. Robber fly resting on a leaf of blowout grass.

mature. A study of this species by B. Elizabeth Horner and J. Mary Taylor indicated that mating occurs following a period of chasing (with either sex in the lead), nuzzling, and various nose-to-anus and nose-to-nose touching, much as in kangaroo rats. The northern grasshopper mice studied by David Ruffer performed very similar sexual behavior.

During the courtship period both sexes vocalize with chirping notes, but the most famous call of both species of grasshopper mice is a long, high-pitched scream, typically uttered from a standing position with the nose lifted and the mouth opened wide. Males interested in females may utter this call often, as frequently as 20 times in a three-hour period; females may also, though rarely, utter the same or a similar call. In northern grasshopper mice, Mark and David Hafner found that individual, sexual, and species-level differences (though apparently not geographic differences) can be detected in this call. Larger animals have deeper voices than smaller ones, and the call may serve as a long-distance signal for maintaining social contact between individuals. Besides providing directional and distance information, the call may provide social information on the calling individual's species, sex, age, and size. It may also facilitate spacing in these widely dispersed rodents, and since males call more often than females, it may especially assist in mate acquisition and serve as territorial proclamation (figure 31).

According to David Ruffer, territories are typically defended by nonviolent agonistic encounters, and scent probably plays a major role in the establishment and maintenance of a territory. Scent is produced by glands in the anal region, and marking is done by rubbing this area in holes dug at the edges of territories. A line of small burrows may thus mark the edge of a well-established territory. When the grasshopper mouse does attack a conspecific or some other species, it chases and repeatedly pounces at its victim until it gains a hold with its forefeet; it then bites through the rear portion of the victim's skull, often killing the animal within ten seconds of initial seizure. Although both male and female grasshopper mice will kill other mice individually, the male assumes the role of attacker when a pair is present. At times a grasshopper mouse may attack a rodent up to three times its own weight, including animals as large as hispid cotton rats. During intraspecific encounters, dominant-subordinate relationships are quickly established, and at least in captivity the subordinate individual is likely to be killed within the first 24 hours of contact. Nonviolent encounters are evidently im-

Fig. 31. Northern grasshopper mouse behavior, including sonogram of adult scream (A) and screaming posture (B), a normal relaxed posture (C), and the nose-to-nose (D) and nose-to-genital (E) courtship postures.

portant in maintaining dominance relationships, however, and are thus more likely to occur than are life-threatening ones after the establishment of a dominance hierarchy.

Although grasshopper mice frequently eat grasshoppers, grasshopper sparrows are even more aptly named. Early "food habit" studies by biologists such as S. D. Judd indicate that during half the year grasshoppers constitute about a quarter of these sparrows' foods; during the four-month summer period, about a third; and during June (the peak of breeding activities), more than half. Studies by Anthony Joern on Arapaho Prairie suggest that grasshopper sparrows can significantly reduce the densities of grasshoppers there. In one case, an estimated 27 percent decline in grasshopper density occurred, with a corresponding reduction in grasshopper species diversity. In controlled studies using captive birds and regulated prey abundance, Joern determined that one of the four grasshopper species made available was selectively preferred as prey, although some switching in prey selection was observed. These birds were found to spend about 50 percent of their time searching for prey, 10 percent in prey handling, and their remaining time in nonforaging activities.

The grasshopper sparrow was actually named not for its primary prey but for the soft, buzzy, insectlike vocalizations that are the male's territorial advertisement "song." Grasshopper sparrows arrive in the Sandhills during early May, and soon thereafter the dry songs of the inconspicuous males can be heard everywhere. The birds sing from a variety of high perches, including fenceposts, within their territory, generally preferring the highest available sites. Typically, the song perches are near the edges of the territory, which average up to about one hectare in area.

In the male's advertisement song a few preliminary "tic" notes generally precede the extended buzzing. A more sustained song, lasting up to five seconds, includes the usual "grasshopper song" as an introduction, followed by a prolonged series of rather melodious notes. Often this song is uttered on the upward phase of a short display flight, after which the bird quickly drops back down into the vegetation and completely disappears. The male's grasshopper song is uttered off and on throughout the day, especially during the incubation period. The much rarer sustained song begins at about the time of the female's arrival on the territory but is largely confined to evening hours. A trilled song that evidently serves as a pair-bonding vocalization is also uttered by the male, and a similar vocalization by the female. Her trilled utterances may serve to let the male know that she is approaching the nest or simply to declare her presence on the territory.

Nests of grasshopper sparrows are even harder to find than are the birds themselves. Nests are well hidden at the bases of plants and often have one or two inconspicuous pathways leading to them. They are built during May, with second nestings or possible renesting efforts extending to early July. The female alone incubates the eggs, probably for about 12 days, though few if any incubation periods have been accurately timed in this little-studied species. During the incubation period the male defends the territory and advertises it. Following hatching he participates fully in food gathering, nest defense, and most other brood-rearing activities except for the actual brooding of nestlings. Second nestings may be fairly common, at least in some parts of the species' range. After the last broods have fledged, the adults and young evidently remain in the general breeding area until the fall migration.

About the same time the grasshopper sparrows are establishing their territories in lowland meadows, the somewhat misnamed upland sandpipers are doing the same. Once even less appropriately called the "upland plover," the upland sandpiper not only is not a true plover but may be described as an "upland" bird only in that unlike most shorebirds it is ecologically wholly independent of shoreline habitats. The native mixed-grass prairies of the central Great Plains are its preferred if mostly vanished home, although its breeding range extends marginally from the interior Alaska grasslands to those of upper New England. But it finds its perfect remaining breeding habitats in the Nebraska Sandhills meadows, where there is an abundant supply of grasshoppers to eat and a rich cover of middle-height grasses for nesting.

Few large hay meadows in the Sandhills are wholly lacking in these graceful birds, whose presence is generally made apparent by their seemingly out-of-place appearance on fenceposts that are often far from water, or the liquid trills of a male performing its song-flight display while gliding high above its territory on down-curved wings. Under perfect conditions the call may carry nearly a mile, and the bird may be so high as to be nearly out of sight—no wonder they have been called "spirit-voiced" birds! At times, however, a displaying bird high in the sky will suddenly terminate its calling and then plunge quickly back to

Fig. 32. Upland sandpiper nesting among plains phlox.

earth. On many other occasions it will call repeatedly while flying on quivering wings only a few yards above the grassy meadows.

When visually advertising their territory or simply scanning the area for possible danger to the nests or young, upland sandpipers are especially likely to stand on elevated sites. Thus, like burrowing owls, grasshopper sparrows, and meadowlarks, they are to be looked for along all roadside fences near hay meadows. As they alight on fenceposts, upland sandpipers invariably hold both wings above their bodies momentarily, ballerinalike, before inserting them delicately inside their flank feathers. Aldo Leopold once wrote, "Whoever invented the word 'grace' must have seen the wing-folding of the plover." Similarly, whoever might be looking for a new definition of exhilaration need only observe the song flight of the upland sandpiper.

Rather than the meadowlark, which is found nearly throughout the continent and is the "official" bird of half a dozen states, the upland sandpiper might have been selected as Nebraska's state bird on the basis of its close association with our native prairies, its understated beauty (figure 32), and its lovely song.

CHAPTER NINE

Brooks and Rivers

Sedges, Sticklebacks, and Swallows

Fig. 33. Male yellow-headed blackbird on cattail leaves.

THE STREAMS OF THE SANDHILLS are sedge-lined. That is, they are festooned with sedges, grasses, and an abundance of other aquatic to semiaquatic organisms characteristic of the region's numerous creeks and small rivers. This life-giving circulatory system begins as innumerable tiny aquatic capillaries, which occasionally burst out unexpectedly from the bases of dunes as artesian springs, then sometimes flow into a Sandhills lake or wander aimlessly as clear, cool brooks through little valleys and wet meadows. These brooks gradually make their way in graceful arcs through nameless Sandhills valleys, usually headed toward the east and southeast, where they eventually reach and merge with the Cedar, Calamus, Dismal, and Loup Rivers, all of which are part of the greater Loup drainage and constitute about half of the Sandhills drainage system. A few streams flow northwardly or northeastwardly to enter the Niobrara River drainage; these represent about a quarter of the total Sandhills drainage network and include the relatively short Snake River and those brooks emerging as cold springs in the springbranch canyons of the Niobrara Valley. The remaining surface drainage from the Sandhills reaches the North Platte, Platte, and Elkhorn River systems (figure 34).

The hay meadows described in chapter 8 merge into lower and wetter sites, which were described by Raymond Pool as the meadow formation. He stated that the rush–sedge association—dominated or characterized by several species of rushes, spikesedges, sedges, and bulrushes—makes up the wettest phase of this formation. Somewhat distinct from the wet meadows, which are usually controlled by bulrushes and grasses surrounding the sunlit marshes, is a separate streamside marsh association that occurs in well-shaded areas. Here, shade-tolerant plants such as spotted touch-me-not, willowherb, and other forbs able to tolerate water-saturated soils are prevalent.

W. L. Tolstead called the lowest part of the shoreline zone around wetlands the "hygrophytic grass and sedge zone." He stated that the width of this zone varies from a few feet to more than 30 feet (9–10 meters), depending on the slope of the land, and the associated plants may be flooded to a depth of several inches (perhaps 15 centimeters) in spring. In the northern Sandhills this periodically flooded zone is usually dominated by bluejoint and Sartwell's sedge. Around cool springs various bulrushes, water plantains, and bur-reed are typical where there are moderate fluctuations of the water table; where the water table

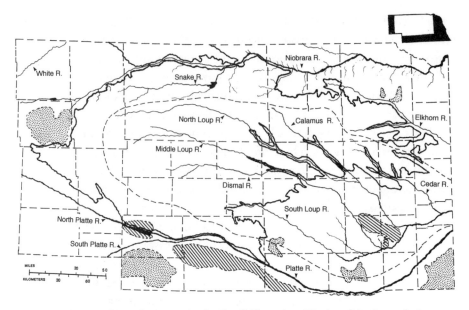

Fig. 34. River drainages in the Sandhills region. Limits of the Loup drainage are indicated by a broken line. Also shown are major historic changes in regional groundwater levels, including losses (stippled) and increases (hatched) of ten feet or more as of 1991 (after a University of Nebraska Conservation and Survey Division map).

remains just under the surface throughout the year, such plants as bottlebrush sedge and mannagrasses are common. The higher and somewhat drier portions of these shoreline zones are dominated by such familiar tallgrass species as big bluestem and Indian-grass, together with alkali cordgrass and many other meadow-adapted grasses.

The cool and clear headwater streams in the Sandhills support several rather rare or at least variously localized species of generally boreal and minnow-sized fishes, including the pearl dace and the brook stickleback. Several brook-adapted fish species, including the brook stickleback, blacknose shiner, and three species of dace (pearl, finescale, and northern redbelly) reach or nearly reach the southern limits of their Great Plains ranges in these cool spring-fed waters of the Sandhills and adjoining areas. The blacknose shiner, which in the late 1800s was said to be one of the most abundant fishes in the state, is currently a highly threatened species limited to a few clear and cool streams in the northern Sandhills and upper Niobrara drainages, such as Brush Creek and Gordon Creek in southern Cherry County. The brook stickleback is largely confined to streams up to about six meters wide having sandy or silt bottoms, mostly in the Platte, Niobrara, and Loup basins, though they may be quite common locally in these areas.

The brook, or "five-spined," stickleback in the Sandhills population is notable for the five spines that occur along the back and probably help deter attacks by predators (figure 35). In some populations of this species the spine number varies, and in western parts of its broad geographic range these spines are shorter than in the eastern areas. It is believed that this reduction in spine length and a generally more streamlined body profile facilitates a more rapid escape into vegetation. Another North American species of stickleback usually has nine spines, and a species that is common in Europe and has been extensively studied by ethologists there has only three. Sticklebacks are all carnivorous and primarily consume a wide variety of insect larvae and small crustaceans. The brook stickleback can tolerate virtually freezing water temperatures during winter, and breeding is largely limited to waters that are no warmer than about 19 degrees Celsius during the breeding season.

Spawning in the brook stickleback occurs early in spring, perhaps under the influence of an increasing photoperiod. At that time the adults, which become sexually mature within a year of hatching, begin to abandon the aggregated schools that characterize the nonbreeding period. Then, lake or pond populations (such as in the Great Lakes region) move from deeper and colder waters into shallower waters, and some river populations may move upstream. In Nebraska, where there are evidently few if any such seasonal upstream movements, males establish territories and begin building nests in March or April.

Adult males are much darker overall than nonbreeding individuals and may become jet black dorsally and on the dorsal and anal fins. Anteriorly, the male may vary from reddish to black. The dark colors of the upperparts are contrastingly variegated with yellow on the sides and exhibit more uniform coppery tones below. Breeding males also have bright yellow eyes, each with a vertical black band extending across the iris. This black eye stripe is most conspicuous during territorial interactions and sexual encounters. Females undergo more limited color changes, passing from a rather uniform pale green nonbreeding color to generally more olive tones when breeding. They also acquire diffuse black to brown mottling or dotting on the upperparts but never develop dark eye stripes comparable to those of males.

The territories established by male sticklebacks center on their nests, which they vigorously defend against others of their own species and similar-sized fish of other species. This defensive behavior persists until the young have left the nest. When two males have closely adjoining territories, their narrow boundaries are readily recognized by both and can rarely be transgressed without stimulating an attack by the resident male. Territories may be quite varied in size, from about 250 to 1,700 square centimeters of substrate area, within which the male usually builds several nests: a half-dozen or more in the larger territories, only two or three in the smallest. In one aquarium study by Howard Reisman and Tom Cade, 14 males constructed a total of 70 nests, and most of the spawning occurred in nests located within the larger territories.

Males construct their nests by attaching roots, twigs, filamentous algae, dead leaves, or other available materials to a solid vertical substrate such as a rock surface or a stem of emergent vegetation. The attachment is done by means of a sticky, mucuslike material produced by the male's kidneys and extruded to the surfaces of the nest. A nest can be largely constructed within a single day, although much repair behavior and addition of materials may go on over a prolonged period, so that it gradually increases in size. The completed nest takes a generally ovoid or globular form up to about the size of a golf ball, with a hole in one side. Nests are situated at varied distances, sometimes as much as 30 centimeters, above the bottom substrate.

While males are building their nests, the females swim about individually or in groups outside the area occupied by the males. Females are sometimes independently attracted to finished nests, but usually the male will actively swim toward any gravid female. He approaches her from above, erecting his spines and rapidly vibrating his caudal and pectoral fins. He may then pummel the female with his snout on her head and flanks, causing her to sink toward the bottom, and attempt to lead her toward the nest by swimming with his spines erected, his body arched, and his mouth often held open. The male will then intently "show" the nest entrance to the approaching female by fanning his tail in an exaggerated manner while positioning his head immediately in front of or even partly inserting it into the opening. If the female enters the nest, she pushes her head through the opposite side, then remains motionless for a time with only her head and tail visible (figure 35). The male approaches from behind and prods her underside with his snout, stimulating her to shed her eggs with accompanying shuddering movements. About 75–100 eggs may thus be laid in the nest.

After laying her eggs, the female quickly exits the nest through the side opposite the entrance. The male then immediately enters, fertilizes the female's eggs without pausing, and leaves through the exit hole made by the female. After that, he may chase her from his territory and then return to tend and protect the nest. He may repair the exit hole and often positions himself in front of the entrance, fanning a current of water through the nest with a strong beating movement of his pectoral fins. At times the male will build a new nest and transfer the eggs to that site. If an egg should fall from the nest he will retrieve it and carry it back in his mouth.

During the week or so before hatching occurs, the male constructs a "nursery" by reorganizing the upper part of the nest so that it forms a loose network of vegetation. As the embryos hatch, they float upward until they are captured by the vegetational mesh; if any pass through, they are quickly retrieved by the male and returned to the nursery. After a day or two the young begin to escape from the nursery so rapidly that the male is unable to catch and return them, and thereafter they are on their own.

Sticklebacks are not alone in constructing globular nests with lateral entrances. Not far above the water, cliff swallows sometimes fill every available vertical surface on concrete culverts and bridges with their adobelike nests of dried mud. The presence of a roughened vertical surface to which wet mud can be firmly attached, a protective overhanging ledge or wall, and a nearby source of claylike mud are the minimum nesting requirements. Cliff swallows are extraordinarily common around Sandhills rivers and brooks wherever such specialized sites can be found. Colonies of several thousand nests have been counted on some bridges in the North Platte Valley; even small culverts under creeks may support a half-dozen or so. Given the right opportunities for nesting and abundant foraging, cliff swallows are among the most colonial of all nonmarine birds in North America.

These swallows usually arrive in the Sandhills in late April and immediately begin to reoccupy the remains of the previous year's nests, which often persist from year to year in the more protected sites. Where these

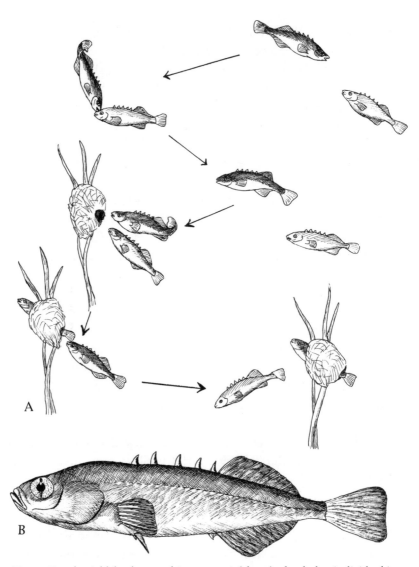

Fig. 35. Brook stickleback courtship sequence (above); the darker individual is the male. The spines are evident in the illustration (below) of a breeding adult male.

no longer exist, or if new areas are being colonized, the nests must be built from scratch. The construction phase requires about two weeks, and both sexes actively participate. To gather their building material, the birds stand and dig out small globs of mud with their short, wide bills, simultaneously raising their wings vertically and quivering them rapidly. This behavior is believed to allow females to gather mud without being sexually assaulted from above by males. Unlike the males of many species, which accompany their mates on excursions from the nest site, cliff swallows must guard their nests to avoid having them taken over by other males. As a result, members of the pair usually take turns remaining at the nest during the construction phase, as well as in the egg-laying and incubation phases, until the time of hatching.

There would seem to be a time-related advantage for swallows in taking over a usable previous year's nest and beginning their egg-laying phase early, but there are also dangers involved: swallow bugs, which are related to bedbugs and bat bugs and suck the blood of nestlings, can survive from year to year in abandoned nests. Swallows therefore tend to avoid old nests that are already heavily infested with bugs. Taking over a relatively uninfested nest, however, does save two weeks of building and also allows the young to be hatched before bug infestations become extremely severe, which occurs in late summer. Swallows are more likely to construct new nests from scratch in larger colonies than in smaller ones, probably because heavier infestations of these parasites occur in the existing nests of these larger colonies.

A decade of research by Charles and Mary Brown on cliff swallow behavior and breeding biology in the North Platte Valley has illuminated the intricacies of behavioral adaptations associated with colonial nesting, which these birds display to a degree unmatched by any other North American passerine, if not any other colonial vertebrate. The Browns soon established that swallow bugs could be the cause of nearly half of all deaths of nestling swallows and of reduced body weight in the surviving young. Coloniality thus exerts a heavy reproductive price for swallows and must be counterbalanced by advantages if it is to be favored. One such advantage discovered by the Browns is that larger colonies have a greater level of group vigilance in detecting nest predators such as bull snakes, even though the birds have no effective defense against them. Since the peripheral nests are those usually taken by such predators, there is a relative reproductive advantage for birds nesting in interior portions of the colony. Generally, however, overall reproductive success was not found to vary significantly with colony size.

One of the Browns' most remarkable findings was that cliff swallows are effective social parasites, the females often laying their eggs in the nests of others in the colony. Cliff swallows normally lay four eggs on each of four consecutive days. The host female can recognize and will discard an alien egg only if it appears at least five days before her own laying begins; if it is deposited within three days of her starting her own clutch, the host female will accept it and incubate it together with her own eggs. Such early laying increases the chances that the parasite's egg will be among the first to hatch and thus increases the chances for that nestling's survival. Even when the parasitic egg is laid after the host's clutch has been begun, it is usually deposited early in the egg-laying period.

Not only do parasitic females lay eggs in other birds' nests, but they may also even move eggs between nests with their bills, transferring one of their own into the nest of a neighbor. Egg transfers may occur at any time in the incubation period and represent a previously unknown method of social parasitism. Parasitic females evidently do not remove host eggs from clutches when they lay in nests already containing some eggs, but some birds—especially males—may visit an unattended nest and either roll an egg out or pick it up and drop it from the nest opening. This seemingly spiteful behavior tends to keep clutch sizes from becoming undesirably large where nest parasitism occurs.

As a result of parasitic behavior, some 20 to 40 percent of cliff swallow nests become parasitized with one or more foreign eggs. Remarkably, not only do most of these parasitically laid eggs hatch (about 75 percent), but those swallow nests containing parasitic eggs actually tend to produce more healthy fledglings than do unparasitized nests. Parasitic females were also found to lay more eggs and produce more fledged young than nonparasitic individuals. Of course, females parasitizing the nests of their neighbors may in turn be parasitized by other birds during their absence from their own nest, so that parasitic individuals may end up tending and feeding nearly as many young as do their victims. Nevertheless, by spreading her eggs around, a female may increase the chance that some of her young will survive, especially given the high mortality rates of nestlings reared in heavily bug-infested nests. The incidence of egg parasitism should therefore increase with the increasing uncertainty of reproduction. Consistent with this prediction, in colonies where the chances of successful reproduction were high, the Browns found fewer nests containing parasitic nestlings.

One clear-cut benefit of coloniality is that cliff swallows have been found to use "information centers" to find the locations of rich food sources such as dense aggregations of aerial insects: individuals that have been unable to find food on a foraging trip will locate a more successful neighbor and follow that individual back to its food source. Moreover, although they have no effective defense against predators such as bull snakes (or, probably, large avian predators), these birds do perform mobbing behavior toward such enemies, or models resembling them. Charles Brown and John Hoogland established that the colonially nesting cliff and bank swallows are less likely to dive at predator models than are the more solitarily nesting barn and rough-winged swallows. Apparently, solitary species have to take greater individual risks in mobbing than do colonial ones, and so one probable benefit of group living for cliff swallows is the reduced risk per capita of being caught and killed by a predator while engaging in mobbing behavior toward it.

Barn swallows build nests that are somewhat similar to those of cliff swallows, but they almost always place their grass-and-mud nests where there is bottom support in addition to a lateral support surface. As a result, barn swallows are mostly found nesting on the beams or eaves of open buildings, rather than on the clifflike sites favored by cliff swallows. In a few places both species will nest and feed in proximity, but barn swallows in the Sandhills are, appropriately enough, most often associated with barns.

When Lake Ogallala was being modified in conjunction with the conversion of Kingsley Dam to attain increased hydroelectric capabilities (see chapter 4), Cedar Point Biological Station lost some of its shoreline. As partial compensation, Central Nebraska Public Power offered to give the station an old building that was situated near the dam and not of immediate use to the utility. Cedar Point accepted the offer and made arrangements to move the building on a flatbed trailer the half-mile or so to the station grounds. As it turned out, two active nests of barn swallows were present in the building's eaves, and I fully expected these nests to be abandoned or destroyed in the course of the move, but that was not the case. As the building slowly lumbered up the narrow road, the parent barn swallows flew excitedly above, screaming and circling overhead. As soon as the structure had been put in place, they settled in to complete their nestings as if nothing had changed. In this way Cedar Point acquired its first pairs of nesting barn swallows, and the building became known thereafter as the "swallow barn."

The birds have used the swallow barn year after year since then. They usually rear at least three broods per summer, in nests placed in each of the four corners of the building, and with a seemingly random rotating use of the available nest sites. A few years ago I was asked to provide some recently abandoned nests to a biologist who was studying the infestations of bird nests by bloodsucking bugs and similar nest fauna, and I decided that I could easily take one of the recently used swallow nests for this purpose. On the afternoon before I was scheduled to leave camp at the end of the session, I climbed up to collect a nest, carefully selecting one located at the opposite end of the building from that currently being used by the swallows. I placed the nest in a plastic sack, pleased that I had apparently not disturbed, or seemingly even been seen by, the nesting pair, and returned to my nearby cabin to continue packing. About fifteen minutes later I heard a scratching noise at one of the cabin windows and was astonished to see a barn swallow hovering there, its wings beating at the pane. I could not have been more amazed. I was even more surprised when, an hour or so later, I was greeted by intense mobbing behavior by both adult swallows as I left the cabin to go to supper at the central dining facility. The birds were still perched near my cabin, watching me intently when I returned, and screamed at me until I disappeared inside. The next morning I fled camp as inconspicuously as possible, feeling almost as guilty as a kidnapper and hoping that the swallows wouldn't pursue me down the road, casting obscenities.

The following spring I was distinctly relieved when the swallows paid me no special notice as I cautiously moved into my cabin. They were back in their usual place, fully engaging in life's struggles with the same intensity as ever, and as much as part of Cedar Point's interacting web of life as the bluestems, the soapweeds, the cedars, and the midges.

CHAPTER TEN

Marshes and Lakes

Whispers, Bells, and Trumpets

Fig. 36. Forster's tern, mating pair.

MOST "LAKES" IN THE SANDHILLS are marshes. At least in an ecological sense, few wetlands there qualify as real lakes, being too shallow to develop much thermal stratification in summer and too small to have the barren shorelines that lakes usually produce through wave action. The larger wetlands are concentrated in the western and north-western parts of the region, beyond the upper limits of the Loup and Calamus drainage system in the northern and central Sandhills, and similarly beyond the limits of Blue and Birdwood Creeks in the western Sandhills. In this latter region, where the overall land-surface gradients are slight and many of the wetlands have been effectively plugged at their bottoms with clay runoff from adjoining tablelands, alkaline and saline wetlands are common. Those in the central and northern part of the Sandhills, having developed over sandy loam or fine sand, are only slightly alkaline. Many of the more easterly lakes were once larger, and their one-time shallow areas now often consist only of wet meadows. Some are probably thousands of years old but are undergoing natural aging (eutrophication) slowly, inasmuch as their nutrient concentrations seem to have remained quite stable over the years in which they have been studied.

As is true of the shorelines of Sandhills brooks and rivers, there is a gradual gradient between the plant communities of the lowland mead-ows and those of the wetlands proper, as species arrange themselves according to their water requirements, flooding tolerance, and various other ecological conditions (figure 37). Jean Novacek has tallied the total number of ecologically associated species of vascular plants present in each of these Sandhills wetland habitat zones as follows: 32 subirrigated meadow species, 27 emergents, 27 semiaquatics, 17 submergents, and 6 floating. Since several of these species occur in more than one habitat zone, the species diversity in any single zone may be greater than these numbers indicate. Novacek has also listed 54 nesting species of birds associated with Sandhills wetlands, including 15 species of waterfowl, 9 shorebirds, 4 swallows, and 3 species each of grebes; herons or bitterns; rails or coots; gulls or terns; and blackbirds or meadowlarks.

Proceeding from the relatively dry hay meadows described in chapter 8, the increasingly aquatic plant communities of the Sandhills were sequentially organized by Raymond Pool into three major subdivisions: the meadow formation, the marsh formation, and the water plant for-mation. The meadow formation, described in chapter 8 as the sub-

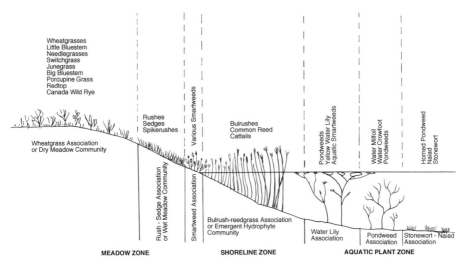

Fig. 37. Profile of wet meadow and aquatic plant communities in the Sand-hills, indicating typical plants of each vegetational zone.

irrigated low meadow or hay meadow association, also includes a fern-dominated wetter variant or "fern meadow association." There is a similarly damp "water hemlock association" (dominated by water hem-lock) and an even wetter "rush–sedge wet meadow association" (men-tioned in chapter 9). The hay meadow association is the best developed and ecologically most important of these several community types, but the others are locally common, and the fern meadow type is best devel-oped in the lake district of Cherry County.

Another fairly rare wetland variant in the Sandhills is the fen. A fen is a boglike wetland with peat accumulations at its bottom, produc-ing a water-saturated layer of partly decomposed organic soil that in Nebraska may be over two meters in thickness. Unlike bogs, fens re-ceive their water directly from groundwater sources, making them much more fertile than the more ecologically isolated and thus more nutrient-poor and acidic bogs. True fens occur in two kinds of ecological situa-tions in the Sandhills: in the valleys associated with stream headwaters, such as at the upper ends of Hay, Big, Boardman, and Minnechaduza Creeks in the northern Sandhills; and at the upper ends of some Sand-hills lakes, where they grade into more typical low meadows. Fen vegeta-tion is similar to that of low meadows but includes quite a number of rare species that appear to be glacial relics of species now mostly found well to the north of the Sandhills. Among these plant rarities are cotton-grass, buckbean, northern bog violet, swamp lousewort, rush aster, marsh marigold, western red lily, closed gentian, and water horsetail. Probably these fens are thousands of years old and began to form during a cooler climatic postglacial period starting about 5,000 years ago.

The most common vegetational representation of the marsh formation in the Sandhills is the bulrush–reedgrass association in which cattails, several species of bulrushes, and common reeds (also known as phragmites) share dominance. This association lines innumerable Sandhills wetlands, and the thick, dense cover of the tallest plants of these shoreline zones provides support for the nests of black-crowned night-herons, bitterns, red-winged and yellow-headed blackbirds, marsh wrens, and swamp sparrows, to mention only a few of the commonly nesting avian associates. This community type also extends out into shallow water, where grebes, coots, rails, terns, Franklin's gulls, and some diving ducks such as ruddy ducks and redheads often place their nests. Nesting over the water certainly helps to reduce predation by animals such as skunks and bull snakes but also exposes the nests to potential damage resulting from wave action and from rapid water-level changes with which the birds may be unable to cope effectively.

Within the marshes and lakes themselves and beyond their emergent borders, the water plant formation holds sway. Floating and submerged pondweeds, water crowfoots, and others make up the pondweed association, which at times almost chokes the shallower, less alkaline, and less turbid Sandhills ponds. The organic debris produced by untold generations of pondweeds accumulates on the bottom, gradually filling in the ponds. In the process these sediments provide a rich source of food for many aquatic invertebrates, and the seeds and leaves of pondweeds, wigeon grass, and other plants of this association are avidly eaten by waterfowl.

In somewhat deeper waters the water lily formation covers large areas, generally occupying a broad zone between the pondweed association and the deeper beds of submerged aquatics. Yellow water lilies may occur in dense single-species stands, or they may intermingle with various floating-leaf species of pondweeds. The shade of these broad-leaved aquatics may be important for aquatic animals, and the large leaves provide escape cover for young broods of waterfowl. Small and entirely floating species such as the duckweeds offer a rich source of waterfowl foods in this shallow-water zone as well as across the entire lake or pond surface, regardless of its depth.

Finally, the sandy bottoms of some lakes and ponds have well-developed carpets of submerged aquatics such as algal stoneworts, and various submerged pondweeds such as naiads and horned pondweed. Compared with the floating-leaf pondweed association, these plants contribute relatively little organic production to the aquatic community and offer aquatic animals fewer opportunities for obtaining food or finding escape cover.

None of the lakes in the Sandhills is more than about 4.3 meters (14 feet) deep, and most are only a meter or so. As a result, they never develop long-term thermal stratification, and water temperatures fluctuate markedly and rapidly. There are correspondingly rapid changes in oxygen content, and probably wave action has an important role in influencing relative oxygen availability. The largest lake complex in the Sandhills covers about 800 hectares (2,000 acres), but the vast majority of more or less permanent lakes are less than 40 hectares.Collectively, the nearly 1,700 lakes studied in a survey of 13 Sandhills counties by Bruce McCarraher in 1977 averaged about 24 hectares and had average maximum depths of less than a meter.

If wetlands smaller than "lakes" are included, the vast majority of the Sandhills wetlands are even smaller than that. The most complete survey, based on remote-sensing data (from the scientific satellite Landsat) obtained in 1979–80 by J. K. Turner and D. C. Rundquist, indicated that the total Sandhills wetlands then amounted to about 526,000 hectares (1.3 million acres), or slightly more than 10 percent of the total Sandhills area (figure 38). About 85.5 percent of this total consisted of fresh meadows, 8.5 percent of open freshwater wetlands (lakes, ponds, and reservoirs), 5 percent of shallow to deep marshes, and about 1 percent of riparian vegetation. The estimate of total marsh area (64,500 acres, or 26,000 hectares) was remarkably close to the 26,750 hectares estimated by the Nebraska Game Commission during the 1960s. Additionally, 84 percent of all the Sandhills wetlands were judged by the Game Commission's survey to be no more than ten acres (four hectares) in area. Estimates of the total area of wet meadows and open freshwater wetlands, however, differed greatly in the two studies.

In the western Sandhills of Garden and Sheridan Counties, where

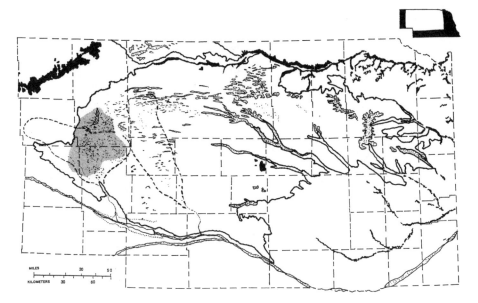

Fig. 38. Distribution of wetlands in the Sandhills region, showing area of poor drainage and highly alkaline wetlands in the western Sandhills (dark stippling), riparian woodlands (light stippling) and other regional forested areas (inked). The blocked drainages of Blue and Birdwood Creeks (dashed lines), and the approximate limits of extinct Lake Diffendal (dotted line) are also indicated.

alkalinity and salinity levels are often extreme, dissolved minerals in the water may prevent virtually all the typical aquatic vegetational communities described above from developing (figure 39). McCarraher's studies of Sandhills lakes in the 1970s indicated that many of these western lakes are highly mineralized, and that 2,000 to 2,200 very small and seasonally intermittent or "playa" lakes are present. About 60 percent of such lakes are, ecologically speaking, alkaline eutrophic and are chemically classified as the sodium bicarbonate type. Stoneworts and pondweeds such as sago pondweed are common aquatic plants in these wetlands, and they are well lined with emergent vegetation such as three-square ("chair-maker's rush") and hardstem bulrush. The remainder are more strongly alkaline lakes of the sodium-potassium hydroxide (infrequently) to carbonate or (usually) bicarbonate types. Although the more alkaline wetlands often have little or no shoreline vegetation, or at most only a saltgrass zone, many of the only moder-

Marshes and Lakes 121

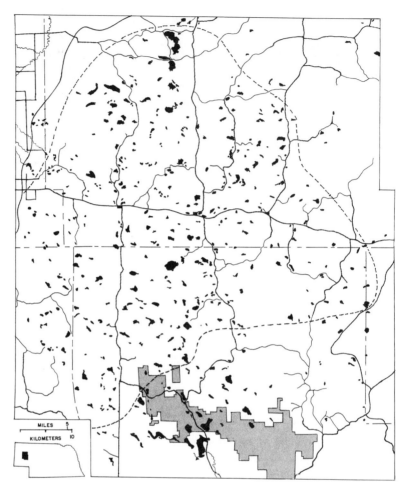

Fig. 39. Wetlands (inked) in the western Sandhills of northern Garden County and southern Sheridan County. Crescent Lake National Wildlife Refuge (shaded) and approximate limits of highly alkaline wetlands (dashed line) are indicated.

ately alkaline lakes produce tremendous seasonal populations of rotifers, cladocerans ("water fleas"), copepods, and various phyllopod ("leaf-footed") shrimp and other crustaceans (figure 40). Such invertebrates frequently provide rich sources of food for aquatic birds. More highly mineralized lakes may have large populations of such strange-looking phyllopod crustaceans as fairy shrimp. The most alkaline "soda" lakes

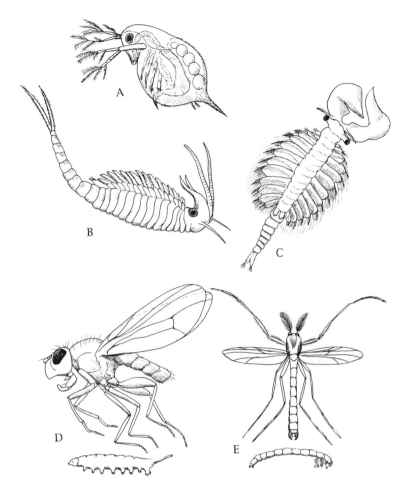

Fig. 40. Common aquatic invertebrates of Sandhills wetlands, including cladoceran water flea *Daphnia* (A), fairy shrimp (B), brine shrimp (C), adult and larval brine fly (D), and adult and larval midge (E).

typically support only the highly salt-tolerant brine shrimp. This widely distributed crustacean has been found in hyperalkaline Sandhills wetlands having pH levels and dissolved concentrations of potassium ions and of carbonates, bicarbonates, and hydroxides even higher than those of the Great Salt Lake and the typical playa lakes of the American Southwest.

The water of the hyperalkaline wetlands is too alkaline for ducklings—and probably most other birds as well—to drink, and no fish can tolerate the highly alkaline waters of bicarbonate lakes, but their use by breeding and migratory shorebirds is often remarkably high. Despite their extreme alkalinity levels (occasionally ranking among the highest of any lakes in the world), these wetlands sometimes display a surprising high productivity, dependent upon their bottom-level or benthos organisms and, especially, algae and other phytoplankton, which are consumed by variously alkaline-tolerant invertebrates such as rotifers, cladocerans, and brine shrimp. The hyperalkaline wetlands generally also have extremely large populations of brine flies, whose larvae feed on algae and detritus and whose adults emerge in vast numbers during the warmer summer and early autumn months. These in turn are avidly eaten by many kinds of wading and swimming shorebirds such as red-necked and Wilson's phalaropes, American avocets, black-necked stilts, willets, and many migratory sandpipers.

Perhaps the most ecologically important insects to be found in the Sandhills marshes and lakes, providing a key link in the food chains of both aquatic and terrestrial animals, are the mosquito-sized midges (figure 40). The adults of these insects form diaphanous and agile clouds in the summer skies over Lake Ogallala and similar uncounted hordes across the wetlands of North America. They live as adults for only a few days; when they die, they sometimes accumulate in windrows along shorelines and around the bases of lakeshore buildings. There are many species of midges. Some lay their eggs in moist soil, under tree bark, or in decaying organic matter, but the majority deposit them in some kind of wetland environment—pond, marsh, river, or lake—on the water surface or just below it. Midges are in fact the most widely distributed of all aquatic insects. Some of their larvae exist at lake depths of nearly 300 meters in the Great Lakes region; others hatched in rainwater ponds can sometimes complete their larval and pupal stages before the ponds dry up. Some species can even survive several years of inactivity as larvae in dry soil, waiting for the water to return so that they can pupate and complete their life cycles.

The deepwater forms of midge larvae are especially rich in the oxygen-carrying pigment hemoglobin, which differs slightly in its chemical structure from the vertebrate version of this important blood pigment. These red larval forms, sometimes called bloodworms because of their hemoglobin-colored bodies, exist in almost incalculable numbers in the oxygen-poor silt, detritus, and other debris of the Sandhills wetland bottoms. John Zimmerman estimated in 1990 that in the Cheyenne Bottoms wetlands of western Kansas there may be as many as 50 bloodworm larvae per square inch, or up to eight per square centimeter, living in the submerged mud flats there, and further reported that the available dry weight of foods from this single source averages about 121,000 pounds (55,000 kilograms) during the nine-month period from March through November. Such a rich supply feeds a great array of shorebirds and waterfowl that probe into the bottom ooze with their bills or suck up this rich organic sludge and extract the larvae from it in various ways. Midge larvae are consumed with equal relish by bottom-foraging fish.

The aquatic larvae of midges feed on a variety of algae and microscopic plant life, or at times on other smaller invertebrates, and themselves are eaten by many larger predatory aquatic insects. Those that swim about rather than burrowing in the mud do so with whipping movements similar to those made by the similar mosquito larvae. Following their four larval instar stages, midges undergo a brief pupal stage before transformation into the winged adult occurs. The pupal stage lasts from as little as a few hours to as long as a few days; when the pupa finally leaves the sludge of the bottom zone and swims up to the water surface, the winged adult almost immediately emerges and flies off, leaving its empty pupal case behind. This emergence of the adults typically occurs about dusk (though in some species, at other times during the day) and may be controlled by light levels, temperature, or both.

Because the adults lack mouthpart structures that would allow them to eat, they begin to starve as soon as they emerge and can live only a few days. During those brief, shining moments the males gather in great whispering clouds that dance erratically above the water and nearby land, especially at dawn and dusk. When the adult females emerge from their pupae, they are immediately ready to mate and soon join the male swarms. The males perhaps recognize them by their distinct flight sounds rather than visually; in any case, every approaching female is instantly grabbed by a male, and the two drop downward in an aerial copulatory embrace. Often, mating is completed before the two reach the ground, but on-ground copulations are known to occur in some species. Then the sexes quickly part, the male immediately returning to the swarm for the balance of his brief life, the female landing and resting for a day or so. Finally, she takes flight one final time and deposits her fertilized eggs as a gelatinous mass on the water surface. Soon thereafter, the female dies. The egg mass swells with water, sinks, and settles on the bottom of the pond, where the embryos begin to develop. When the larvae hatch, they eat their way through the surrounding materials and then are free to burrow into the bottom ooze to begin eating, or perhaps to be eaten themselves.

Among the many bird groups likely to make midge larvae a significant part of their diet are the grebes. These somewhat ducklike birds, often called "helldivers," are so wonderfully adapted for diving that their legs have been shifted far to the rear. In that position they can be splayed out laterally and may be paddled synchronously or alternately, allowing the bird to dive and otherwise maneuver while it forages well below the water surface. Probably the most common and widespread breeding grebe on Sandhills lakes and marshes is the pied-billed grebe, a solitary species that inhabits overgrown areas and is not easily seen. Even in full view the pied-bill is drably and concealingly colored (like rails, it probably finds auditory signals more effective than visual ones in communicating socially within its weed-lined world) and not an especially beautiful member of the grebe assemblage. Far more spectacular are the larger, open-water species such as the eared and western grebes, which bear variously colored ornamental plumes or crests during their breeding periods.

The eared grebe is a widespread species that also occurs commonly in the Old World, where it is often called the black-necked grebe. It has a tapering and pointed but still relatively stubby and somewhat flattened bill that is adapted for capturing both vertebrate and invertebrate prey; however, insects and their aquatic larvae, including midge larvae, are the predominant food source during the breeding season. During that season both sexes have golden yellow ear tufts, surrounded by a jet black head and neck; ruby red eyes are set like rare jewels in this dark matrix.

Eared grebes choose as their nesting areas those marshes that offer a combination of open, rich feeding areas and fairly close sheltered locations. Dense to semi-open stands of bulrushes or similar sturdy-stalked emergent aquatic plants are often selected; an abundance of submerged aquatic plants also provides convenient nest materials. Sites used for nesting range from as shallow as a few centimeters to more than a meter deep, usually a depth that allows for the construction of semifloating nests that are not in great danger either of being tipped over by wave action or of becoming stranded on land by receding water levels. Unlike most other grebes, the eared grebe is distinctly colonial in its nesting, and in some very favorable breeding areas the nests of adjacent pairs may almost or actually touch. As a result of such adaptations for crowding, no real territorial defense has been observed. (In pied-billed grebes, by contrast, nests are well separated, and territoriality appears to be well developed.)

Not only is territoriality undeveloped in eared grebes, but permanent year-to-year pair-bonding or remating with the previous year's mate is evidently lacking. Arriving on their breeding areas in late April or early May, the birds begin to gather around nesting areas, but little or no sign of pair-bonding is evident until only a few days before nest building begins. Nests are built by both sexes, and a recognizable nest can be completed within a day, although materials may be added to it later. Sometimes eggs are laid on rather flimsy and incomplete nests, but these are subsequently abandoned. Mating is done on the nest platform, rather than in the water as might be expected, and must be very difficult for the birds because of the extreme rearward and lateral positioning of their legs, and their resulting awkwardness in standing or walking.

During the nest-building period, pair-bonding behavior is performed—usually, like most grebe displays, in the same or nearly the same way by both sexes, often simultaneously. Several displays are common, but the most frequent one of unmated birds is an advertisement posture and accompanying call, uttered with the crest erect and the neck somewhat stretched. This is frequently the first component of a complex discovery ceremony. One bird utters the advertising call as the other approaches and dives. The first bird, facing the point where the other has disappeared, assumes a raised-wing "cat display." The diving bird may partially emerge several times, usually in a so-called "bouncy posture," but finally fully emerges near its waiting partner in an erect "ghostly penguin" posture: turning away from the other individual, it rises with its neck maximally stretched and its bill held horizontally. Its stationary partner then lowers its expanded wings and the two may perform a "penguin dance," facing each other and paddling in place while remaining erect. The two birds then often "barge" together: that is, remaining in a similar erect, neck-stretched posture, they swim side by side across the water surface (figure 41). They may then finally dive in unison. This complex sequence generally occurs only in the early stages

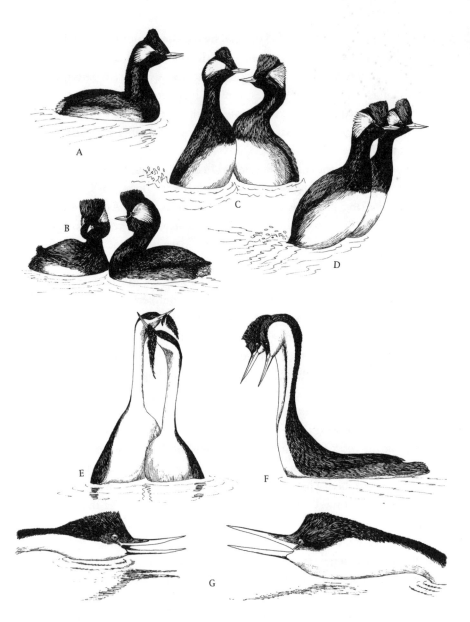

Fig. 41. Grebe courtship displays, including the eared grebe's male advertising posture (A), mutual head-turning (B), penguin dance (C), and barging (D) plus the western grebe's weed trick (E), high arch (F), and ratchet pointing (G).

of pair-bonding, and later versions become variously simplified. Sometimes one of the two birds, in a so-called "weed-trick" display, will gather an aquatic weed and drop it on the water in front of its partner, or simply throw it over the other's back.

On deeper and larger Sandhills lakes that have an adequate supply of the small fish that are almost its entire diet, the elegant western grebe sometimes occurs. This is the largest or at least longest of the North American grebes, and its common vernacular name "swan grebe" gives a good idea of the bird's graceful beauty. Breeding adults are pure white below and blackish gray above. They have long, tapering bills resembling stilettos, and their two-tone head plumage includes a black crown and a short, cropped forehead crest which, when erected in display, somewhat resembles a three-cornered cap. Their eyes during breeding are just as intensely red as those of eared grebes.

Also like eared grebes, western grebes nest colonially, with nests sometimes numbering in the hundreds. They choose emergent vegetation similar in depth and density to that selected by eared grebes, and like those of eared grebes their nests are at times placed extremely close together. Courtship and nest-building activities begin soon after the birds arrive in early May, before the start of actual nesting. Most or all of their displays are performed mutually by both sexes. These include crest raising as the birds swim together; "ratchet pointing," or calling and simultaneously extending their necks and heads toward each other just above the water; and a stereotyped preening behavior called "bob-preening." Another common display is the "high arch," with both birds arching their necks while pointing their bills downward. Equally attractive is the weed ceremony, during which one or both birds dive to bring up bits of aquatic vegetation; then they approach each other, rising in the water and paddling about with bills raised before finally dropping their weeds and engaging in mutual bob-preening (figure 41).

The most interesting courtship display of all is the "race," during which two birds suddenly rise up and rush side by side across the water. Racing, a more rapid and exciting version of the eared grebe's barging, often follows a sequence of ratchet pointing or alternated bill dips and head shakes. The wings are lifted but only partly extended, and the neck is variously bent so that the bill remains nearly horizontal, as the birds skim side by side over the water like competing miniature speedboats (figure 42). The race display is typically performed by a male and female,

Fig. 42. Western grebe pair racing.

yet it is sometimes done by two males or, less frequently, by two or three males and a female or even (though rarely) by a lone female. It ends with all participants diving, and other displays typically follow as the birds emerge.

Watching courting western grebes is something like attending a performance of the Bolshoi Ballet without having had to stand in line for tickets. Even the music is delightful: someone once described the calls of courting western grebes as resembling the tinkling of silver bells, and that is an apt comparison. Seeing and hearing western grebes in spring is thus an experience that all persons who are not absolutely convinced that they will be going to heaven should make every effort to achieve.

If western grebes are the most graceful water birds to be found in the Sandhills lakes, then trumpeter swans must certainly be the most regal. These breathtakingly beautiful swans once nested in the Sandhills, but before the end of the nineteenth century almost all the nesting populations had been extirpated from areas south of Canada. During the early 1960s, however, young trumpeter swans were imported and released into Lacreek National Wildlife Refuge in South Dakota, not far from the Nebraska border. Subsequently, these birds were observed during the breeding season at Valentine National Wildlife Refuge in 1966, and the first modern nesting record for Nebraska occurred in 1968 in Sheridan County. In 1974 some young birds were released at Crescent Lake National Wildlife Refuge, and during the 1970s nesting efforts regularly occurred at Valentine, though with very limited success. But during aerial surveys in 1981 a total of 44 adults and 8 broods were seen in Cherry, Grant, Sheridan, Brown, and Garden Counties. The 23 lakes known to have been used by trumpeter swans during the breeding season ranged from 30 to 625 acres (12–253 hectares), according to James Ducey, with an average size of 178 acres (72 hectares). Nearly all these lakes were predominantly open water; emergent marsh vegetation averaged only about 25 percent of the surface area. Lakes having only slight alkalinity may also have been favored by the birds (57 percent of reported usage) over those of higher alkalinity.

Trumpeter swans move into breeding marshes as early in the spring as possible, since their very long breeding period demands that they not waste any time if they are to rear and fledge their young before the fall freeze-up. In fact, a single breeding effort may require nearly half a year, since nest building may take a week, egg-laying about ten days (five eggs, one laid every other day), incubation 30–35 days, and the fledging of the young 100–120 days. This means that if eggs are laid by the first of May, the young will probably fledge about five and a half months later, or in mid-October. By then there is the threat not only of frost but of the hunting season, for newly fledged and inexperienced swans are in some danger of being mistaken for the ducks and geese that may be legally shot.

Like other swans, trumpeter swans pair indefinitely; that is, the pair bond lasts throughout the lifetime of the mate, assuming that the birds are able to remain together throughout the year. Pair bonds and family bonds are very strong in swans: it is likely that only the premature death of one of the pair would disrupt a normal mating association, and young birds typically remain with their parents for much of their approximate two-year period of sexual immaturity.

Because of their enormous body size (the adults sometimes exceeding 15 kilograms) and foraging needs, the breeding territories of trumpeter swans are also very large. The birds generally will not tolerate another nesting pair of swans on the same lake, unless they remain wholly out of sight. The wonderfully loud and harmonically rich calls of the trumpeter swan, which are resonated and amplified by an extraordinarily long and convoluted windpipe, carry great distances and probably serve as territorial signals. A trumpeter swan's calls emanating from the morning mists in the Sandhills is as stirring today as the sounding trumpets must have been to the Hebrews at Jericho. And admission to the concert, as to the courtship displays of western grebes, is free.

PART THREE

Purviews and Prospects
The Once and Future Sandhills

CHAPTER ELEVEN

Dreams Built on Sand

Cattle, Kincaiders, and Corn

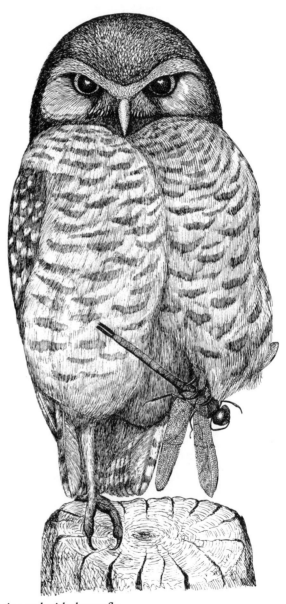

Fig. 43. Burrowing owl with dragonfly.

A HOUSE BUILT ON SAND is unstable. This harsh lesson has been relearned over the past century, since the Sandhills began to be settled in the late 1800s. As human settlement made its impact on the native ecosystems—largely through the loss of the large native grazing ungulates such as bison, elk, and antelope and their replacement by cattle—the "natural" history of the Sandhills gradually became intertwined with their human history. This chapter focuses on human activities and their effects during the past hundred years. The final chapter looks at the present-day ecological, economic, and aesthetic values of the Sandhills and tries to balance them.

The early development of ranching in western Nebraska was facilitated by the opening of Nebraska Territory for settlement with the 1854 passage of the Kansas-Nebraska Act. Ranching activity initially developed around the western and southern edges of the Sandhills, including the environs of Fort Robinson, where it supported the army garrison stationed there. Cattle operations on the High Plains gradually increased until the 1880s, when a series of bitter winters wiped out many herds, and the speculative investments in cattle ranching in the West came to a rapid close. During a late winter blizzard in 1879, however, a herd of about 8,000 cattle that had been grazing along the Niobrara River broke loose and wandered south into the nearby northern Sandhills. The cowboys later sent to find and retrieve any survivors of the blizzard discovered that the lost cattle had not only survived but actually prospered there. Indeed, they found about 1,000 additional head of unbranded cattle that had apparently spent years in the protected valleys of this seemingly barren region. Thereafter, the Sandhills were recognized as potentially valuable grazing country.

This recognition, although welcomed by humans, was a recipe for disaster for the large native grazing animals still using the Sandhills. The elk, which evidently were never common on the Nebraska prairies, had probably been eliminated from western Nebraska by the end of the Civil War. The large herds of bison that had at least seasonally occupied the Sandhills (summering in the Sandhills and perhaps overwintering there or in the Platte Valley) were eliminated by the 1880s. White-tailed and mule deer as well as pronghorns had been largely eliminated as a result of hunting by the early 1900s; later, however, through effective protection and restocking programs, all three species were able to reinvade at least some of the Sandhills region.

The passage of the Pre-emption Act of 1841 and the Homestead Act of 1862 had provided the impetus for the initial phase of homesteading in Nebraska during the Civil War period, and an 1872 amendment allowed Civil War veterans to deduct their time in service when fulfilling the five-year residency obligations. The Timber Culture Act of 1873 subsequently encouraged the planting of trees by permitting eight years of timber culture to substitute for homesteading as the requirement for ownership. But these acts limited land acquisition to 160 acres (65 hectares) per household, far too little to permit profitable ranching operations in the Sandhills. Some enterprising ranchers began to exert self-assumed property rights by grazing their cattle and fencing lands in almost any areas where such actions were not actively contested by their neighbors. They also often bought out tracts of land that had been entered as homestead claims by their cowhand employees, providing the funds needed to purchase the homestead (at $1.50 per acre) after the claimant had lived on it for at least six months. Similarly, lands initially entered as timber-culture tracts were sometimes simply reserved by ranchers who had no intention of growing trees but were only holding the tracts long enough to forestall or prevent their homesteading settlement by others.

In 1904, Congress passed the Kincaid Act, which allowed for the homesteading of claims of 640 acres (260 hectares) in 37 western and southwestern Nebraska counties, including all of the Sandhills counties. The Kincaid Act, together with better federal protection against the illegal fencing of nonhomesteaded lands, stimulated a rapid development of western Nebraska, even though a 640-acre claim was still too small for effective ranching, and for farming, 640 acres of submarginal land offered little improvement over 160 acres. Nevertheless, the Kincaid Act produced a rapid homsteading population influx into the Sandhills, which passed its peak within two decades: most of these homesteading efforts had already failed by the 1920s. With the unusually hot and dry summers and bitterly frigid winters of the 1930s, many of the remaining hard-pressed settlers went bankrupt or abandoned their efforts and sold out to larger landholders. The population of the Kincaid counties thereby entered a long-term decline that continues to the present day. For example, the populations of 13 Sandhills counties that reached their peak of about 41,000 people around 1920 had declined by 1990 to about half that number, or less than 21,000. As the less fortunate landowners were increasingly forced to sell their holdings, the remaining more successful ranchers gradually increased their own; by the late 1900s average ranch sizes in the Sandhills counties had reached about 5,000 acres (2,000 hectares), and many present-day ranches are far larger, with a maximum of about 150,000 acres (60,000 hectares). Between 1980 and 1990 the population loss of 15 predominantly Sandhills-bounded counties of Nebraska has averaged 12–14 percent per decade, or more than 1 percent annually. This approximate rate of decline has held for the entire 70-year period since the 1920s (table 5). The human population of these counties now averages only about one person per square mile (2.6 square kilometers), a density that represents frontier conditions.

While the Sandhills region was being developed as ranchland, the surrounding areas were being simultaneously developed as farmland (figure 44) as irrigation waters became increasingly available to supplement the meager annual rainfalls of central and western Nebraska. The stage for this development was set in 1902 with the passage of the federal Reclamation Act, which provided money for developing water projects that would result in large-scale irrigation in 17 western states. In Nebraska the first of these projects involved Wyoming's North Platte River, which helped to supply irrigation water to about 69,000 acres (28,000 hectares) in eastern Wyoming and 260,000 acres (105,000 hectares) in western Nebraska. The Pathfinder Dam, completed in 1909 by the Bureau of Reclamation, was the first of many Platte River dams and reservoirs; its success resulted in a subsequent series of smaller Bureau of Reclamation dams within Nebraska. In addition, the bureau planned but never constructed such projects as the Nordan Dam on the Niobrara River and the Mid-State Reclamation Project in the central Platte Valley. Both of these ill-planned proposals would have caused enormous ecological damage to the state's most important river systems, and their costs would have far outweighed their potential irrigation benefits. Luckily, a combination of grassroots opposition and financial constraints in Congress prevented their being fully funded.

The U.S. Army Corps of Engineers, more concerned with flood protection than irrigation, entered the dam-building business early in this century, especially on the flood-prone Missouri River. From eastern Montana to northeastern Nebraska the corps planned and built a series of very large, expensive dams that have (except in 1993) effectively con-

Table 5 Demographic and economic characteristics
of the Sandhills counties and Nebraska

1. Counties lying 90–100% within the Sandhills region: Arthur, Grant,
   Hooker, McPherson, and Thomas
   Total area, 3,778 sq. mi. (9,785 km²)
   Total population, 2,628 (0.69 per sq. mi.)
   Total households, 1,500 (0.39 per sq. mi.)
   Total farms and ranches, 500 (0.13 per sq. mi.)
   Total cattle, 135,759 (271 per unit)
   Total irrigated acres, 29,000 (58 acres per unit)
   Median family income (1989), $21,688
   Average population change 1980–90, −12.5%
   Average population change 1930–80, −42.2%
2. Counties lying 75–90% within the Sandhills region: Blaine, Brown,
   Cherry, Garfield, Loup, and Wheeler
   Total area, 9,606 sq. mi. (2,488 km²)
   Total population, 14,411 (1.5 per sq. mi.)
   Total households, 4,500 (0.47 per sq. mi.)
   Total farms and ranches, 2,251 (0.23 per sq. mi.)
   Total cattle, 557,833 (248 per unit)
   Total irrigated acres, 232,000 (103 acres per unit)
   Median family income (1989), $22,718
   Average population change 1980–90, −14.3%
   Average population change 1930–80, −36.5%
3. Counties lying 50–75% within the Sandhills region: Garden, Logan, Rock,
   and Sheridan
   Total area, 5,707 sq. mi. (14,781 km²)
   Total population, 12,107 (2.12 per sq. mi.)
   Total households, 5,700 (1.0 per sq. mi.)
   Total farms and ranches, 1,516 (0.26 per sq. mi.)
   Total cattle, 257,593 (170 per unit)
   Total irrigated acres, 187,000 (123 acres per unit)
   Median family income (1989), $23,245
   Average population change 1980–90, −12.2%
   Average population change 1930–80, −35.4%
4. Counties lying 25–50% within the Sandhills region: Keith, Lincoln, and
   Morrill
   Total area, 4,969 sq. mi. (12,870 km²)
   Total population, 46,515 (9.36 per sq. mi.)
   Total households, 18,100 (3.6 per sq. mi.)
   Total farms and ranches, 2,067 (0.41 per sq. mi.)
   Total cattle, 310,353 (150 per unit)
   Total irrigated acres, 612,000 (296 acres per unit)

Table 5 *Continued*

Median family income (1989), $27,478
Average population change 1980–90, −10.0%
Average population change 1930–80, +22.7%

5. Counties lying 5–25% within the Sandhills region: Antelope, Boone, Custer, and Greeley

Total area, 4,687 sq. mi. (12,139 km²)
Total population, 29,908 (6.38 per sq. mi.)
Total households, 11,900 (2.53 per sq. mi.)
Total farms and ranches, 3,858 (0.82 per sq. mi.)
Total cattle, 436,966 (113 per unit)
Total irrigated acres, 1,204,000 (312 acres per unit)
Median family income (1989), $24,158
Average population change 1980–90, −10.7%
Average population change 1930–80, −48.3%

6. Peripheral Sandhills counties: Box Butte, Dawson, Keya Paha, and Valley

Total area, 3,395 sq. mi. (8,793 km²)
Total population, 32,268 (11.69 per sq. mi.)
Total households, 16,500 (4.86 per sq. mi.)
Total farms and ranches, 1,498 (0.44 per sq. mi.)
Total cattle, 413,761 (276 per unit)
Total irrigated acres, 1,617,000 (1,079 acres per unit)
Median family income (1989), $25,777
Average population change 1980–90, −10.9%
Average population change 1930–80, +1.1%

7. Nebraska overall

Total area, 77,227 sq. mi. (200,018 km²)
1990 population, 1,574,000 (20 per sq. mi.)
Total households, 609,700 (7.9 per sq. mi.)
Total farms and ranches, 60,502 (0.78 per sq. mi.)
Total cattle, 5,838,806 (96 per unit)
Total irrigated acres, 8,000,000 (132 acres per unit)
Median family income (1989), $31,634 (U.S. average, $35,225)
Average population change 1980–90, +0.5%
Average population change 1930–80, +13.9%

Sources: Economic estimates (mostly for 1987, irrigated acreage as of 1984) based on information in Rand McNally's *Commercial Atlas and Marketing Guide,* 120th ed. (1989), and *Nebraska Statistical Handbook* (1991). Recent population and family income data are from 1990 census.

One sq. mi. = 2.59 km²

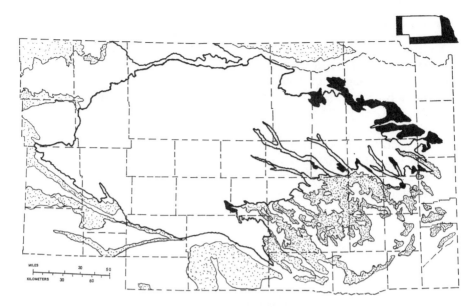

Fig. 44. Distribution of agricultural lands within the Sandhills region (inked), and ranching lands outside the Sandhills (stippled). Based on University of Nebraska Conservation and Survey Division land-use map.

trolled the Missouri's flow over most of its length, generated hydroelectric power, and maintained river conditions suitable for barge traffic. The engineers also constructed the Harlan County Dam on the Republican River, which has provided irrigation water for southern Nebraska and northern Kansas, but their primary purpose has always been flood control.

The amount of land irrigated in Nebraska has so greatly increased during this century that the state now ranks second in the nation (behind California) in total irrigated acreage, and Nebraska constitutes nearly half of the total land area now irrigated by the High Plains aquifer. In 1900 an estimated 150,000 acres (more than 60,000 hectares) was under irrigation in the state, which increased to about 450,000 acres by 1920, about 540,000 by 1940, 2.5 million by 1960, and 7.2 million by 1980. By 1990 an estimated 8 million acres (3.24 million hectares) were being irrigated in Nebraska, an area roughly equivalent to three-fourths of the entire Sandhills region, and about half of Nebraska's total land area under cultivation. Not only has the state's dependence on irrigation water vastly increased, but this water has increasingly been used for growing corn, which demands much more water than do wheat and other grain crops.

During the first half of the century the majority of the water for irrigation came from surface supplies, especially the North Platte and Platte Rivers. It soon became apparent, however, that far more water was available under the ground than above it. Simply tapping into the seemingly inexhaustible Ogallala aquifer opened up a far greater area to irrigation, which had previously been limited largely to major river-basin areas that could conveniently be supplied by river-based water sources.

Pumping devices for irrigation needs became available about the turn of the century; the first of these were quite inefficient, but the development of the turbine-based pump during the first decade of the 1900s was an important breakthrough in irrigation technology. Correspondingly, improvements in internal combustion engines and associated diesel engines gradually improved pumping efficiency and reduced costs. The spreading of electricity to rural homesteads through the Rural Electrification Administration (REA) also made this power source increasingly available.

Next came the technology needed for a new water-delivery approach: the development of sufficient pumping power to produce a sprinkler type of overhead irrigation, rather than a gravity-based and ground-level method. Overheard sprinkling systems suitable for irrigating rather large fields became increasingly available after World War II. These variously involved rotating overhead booms, mobile tow-line systems that could be easily moved from one location to another with tractors, and mobile volume-gun systems. The last could spread water several hundred feet on either side of a pumping device as it was pulled slowly through a field by a cable-and-winch arrangement; however, the volume-gun method required very high water pressures to be effective in large-scale operations, and the devices were very expensive to install and operate. Thus, the invention of center-pivot irrigation in the late 1940s was of special significance to irrigators generally, and it was the first and only system that could have any application for ranchers and farmers in the Sandhills.

The center-pivot system, invented in eastern Colorado, involves a central tower from which a long pipeline extends, suspended above the ground by several large wheels. A complex propulsion system uses water from the main pipeline to force a piston in a hydraulic water cylinder to advance lugs that slowly turn the supporting wheel. Each of the spaced wheels on the pipeline rotates at a different speed: the outermost rotates the fastest and serves as the "master" speed control; the more inwardly located wheels rotate at progressively slower rates.

Despite its considerable mechanical complexities, an early five-tower pivot system was able to irrigate a circular area of 40 acres (16 hectares) very effectively, and it could be raised high enough above the ground to accommodate tall crops such as corn. During the 1950s the patent rights for the center-pivot system were bought by a small company in eastern Nebraska (Valmont Industries, in Valley, Douglas County), which improved the general design to produce extra irrigation acreage potential at section corners, adding about 20–24 acres (8–10 hectares) of additional coverage. The company also began to incorporate fertilizers and pesticides into the water supply, a simultaneous application technique called "chemigation." Such combination methods are often very cost-effective, but great care must be taken to avoid backflow of the chemicals, which would contaminate the original well source. Most modern center-pivot designs now irrigate about 130 acres of a 160-acre quarter-section of land (about 53 of 65 hectares), or 80 percent. The remaining fallow acres can be classified as government-subsidized "set-aside" acreage, allowing the landowner to reap additional direct financial benefits. Inasmuch as a single person can operate and maintain a dozen or more such systems simultaneously, labor costs are relatively low. Some center-pivots a half-mile in length have been marketed, but the smaller designs have proved more efficient and are more popular.

The acceptance of the center-pivot method by Nebraska (and other American) farmers has been remarkable. In 1972 about 2,700 pivot systems were estimated to be present in the state; by 1988 the count was 27,600—a tenfold increase. Most center-pivot sites are located in eastern and northeastern Nebraska, especially just beyond the Sandhills' eastern limits in such counties as Brown, Rock, and Wheeler. But once it was found that the center-pivot device can operate fairly well over somewhat hilly land (including slopes up to a 22 percent grade) and on moderately sandy soils, an eventual introduction into the Sandhills themselves became almost inevitable. Fuel costs average about the same as for irrigating comparable areas by tow-line or overhead-boom methods, and associated labor costs are much lower. Additionally, the level of control over the amount of water applied provides another advantage of adapting center-pivots to sandy soils, which have very limited water-holding capacities. Since no more water is applied than necessary, total water use with the center-pivot method is about the same as by the gravity-based method, yet crop yields may be significantly greater.

During the three decades from the general introduction of the center-pivot system into northeastern Nebraska in the early 1950s to the early 1980s, nearly 200,000 acres (81,000 hectares) came under irrigation in Brown, Rock, and Wheeler Counties. A corresponding increase in well-digging activities in the Sandhills paralleled the introduction of the center-pivot system. The number of irrigation wells increased very slowly from the 1930s to the 1950s but accelerated in the late 1950s, and starting in the mid-1960s it increased even more rapidly. In 24 counties located in and immediately around the Sandhills there were about 75 irrigation wells in 1950, about 220 in 1960, and about 475 in 1970. Similarly, land under irrigation in the Sandhills totaled about 10,000 acres (4,000 hectares) in 1950, 25,000 (10,000 hectares) in 1960, and 50,000 (20,000 hectares) in 1970. By the early 1980s, according to a 1984 estimate by Derrel L. Martin, an approximate 1.25 million acres (over 500 thousand hectares) of sandy soils were being irrigated with center-pivots, about 16 percent of the total area in the state that was under irrigation at that time.

During the maximum growth period for center-pivot irrigation from 1970 to 1981, the state's total irrigated acreage from all types of water-supply systems often increased by 250–500 thousand acres per year and had reached 7.5 million acres (some 3 million hectares) as of 1981. By 1988 the number of center-pivot irrigation systems in Nebraska approached 30,000, and these systems were then irrigating about 3.4 million acres of a total 7.9 million irrigated acres, or about 43 percent of the state's total irrigated acreage. The rates of irrigation and well-drilling activities have tended to level off during the 1980s, as ecological constraints and economic pressures such as rising fuel costs have increasingly come into play.

In a case study, Stephen Hoesel examined the history of center-pivot irrigation in Brown County, at the eastern edges of the Sandhills, one of the first Sandhills counties to be affected by the center-pivot system. Hoesel's analysis provides an instructive model for the effects of this technology on the region generally. In 1955 the first demonstration system was installed on an area of sandy soil near Page, in eastern Holt County. This test was later supplemented by installing center-pivots on cheap ($40–50/acre) sandy lands near Atkinson, also in Holt County. Corn produced on these sites averaged 150 bushels per acre and thus proved the economic value of the technique under marginal soil and difficult topographic conditions—but the ecological price was sometimes high. The center-pivot systems were initially concentrated in the gently undulating eastern Sandhills valley areas, which were often wide enough to accommodate the typical pivot coverage area of 130 acres. When the valleys were too narrow, however, land-leveling operations were used to cut down dune ridges and fill lower places. And once the thin Sandhills soil is cut, the loose, low-nutrient sand that lies below is exposed to massive wind erosion.

During the late 1960s the center-pivot method was adopted widely around the periphery of the eastern Sandhills, and by 1972 Brown County alone had an estimated 87 systems in operation. By 1973 about 550 center-pivot systems had been installed in the 25-county Sandhills region, the great majority in Rock, Brown, Cherry, Holt, Wheeler, and Blaine Counties. At that time there were still no center-pivots established in Hooker, Thomas, or Arthur Counties, and only one in Grant County.

By the late 1980s, with well over 27,000 (active or inactive) center-pivots present in the state, an estimated 3–4 percent of the surface area of the Sandhills region was under irrigation, and all the Sandhills counties had some center-pivots. Although these included more than 9,000 pivots in the 25 counties encompassing the Sandhills, only about 2,700 of them were located within the limits of the Sandhills themselves, the largest concentrations in Rock (13 percent), Wheeler (13 percent), Holt (12 percent), Antelope (8 percent), Cherry (6 percent) and Brown (6 percent) Counties. The central Sandhills area of Hooker, Thomas, Arthur, and Grant Counties had a grand total of only 61 center-pivots, representing about 2 percent of the Sandhills total.

The highest degree of irrigation development occurred in the northeastern Sandhills, especially in Rock, Wheeler, and Holt Counties. For example, Wheeler County had about 60 operating center-pivots in the early 1970s, but within a decade it had 14 percent of its total land area under irrigation, and by 1988 there were 432 center-pivots present. Sig-

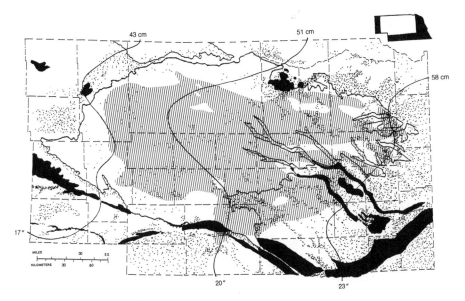

Fig. 45. Distribution of surface-based and groundwater irrigation areas, including major (inked) and minor (stippled) development locations. Also shown (hatched) are the areas having a saturated aquifer thickness of more than 500 feet (150 meters) and average annual precipitation (in inches, 1951–80) totals in the Sandhills region (one inch = 2.54 cm).

nificant reductions in groundwater levels have occurred in Dawes (with 762 pivots), Custer (776 pivots), Antelope (1,462 pivots) and Holt (1,972 pivots) Counties (see figure 45). Generally, the chronological pattern of center-pivot incursion into the Sandhills has been from east to west, gradually progressing across Rock, Brown, and Cherry Counties in the north, and across Wheeler, Garfield, Valley, Loup, and Blaine Counties farther south.

Most of the center-pivot systems installed in Brown County have been used for corn production, and most of the associated wells are about 70 meters deep. In Brown County the best water source is some 150–200 meters down in the Ogallala aquifer, which is easily able to provide the 300–600 gallons per minute (0.02–0.04 cubic meters per second) needed by a typical 130-acre center-pivot system. Initial investment costs (well, pump, motor, and sprinkler) were typically $25,000 to $30,000 in the early 1970s, excluding any land-leveling costs and the value of the undeveloped land itself (often $70–100 per acre). However, after pivot installation, the "developed" land value could immediately rise to as high as $500 per acre, producing a potential land-value profit for the owner even before any actual crops had been raised. Based on average 1970 corn prices of $1.25 per bushel and a countywide average production of 114 bushels of grain corn per acre, the potential before-expenses monetary return would be about $142 per acre. According to Hoesel's calculations, a single 130-acre center-pivot system might thus return a before-expenses income of about $18,500 in the first year of operation. The average price of corn received by farmers throughout the entire decade of the 1970s was $2.04 per bushel; during the 1980s it averaged $2.20; and in the early 1990s it was averaging $2.25–$2.40.

At a 1984 Water Resources Seminar at the University of Nebraska, Roland Langemeier reported that he had placed 11 center-pivots on 9,000 acres of marginal Sandhills land he had purchased in Boone County at $220 per acre and that he was subsequently able to produce corn at an average rate of 111 bushels per acre. The Soil Conservation Service estimated that 50 percent of his developed land consisted of blowouts, compared with about 2 percent of the land under normal Sandhills conditions. Yet Langemeier asserted that "when the sand blows, it's not affecting anyone" and suggested that "we are overemphasizing the importance of one hill in the Sandhills being in one place or the other."

During this same period, Sandhills ranchers found that they could increase their profits by using the corn they had raised with center-pivot irrigation to feed their own cattle. It has been suggested that two center-pivot systems might allow a rancher with 100 brood cows to increase his herd by a third, and that enough winter food could thereby be produced to avoid requiring winter access to hay meadows. Using only one center-pivot system, a rancher might raise enough food to bring 400 calves to a "finished" marketable weight of about 1,100 pounds, increasing each calf's value by about 50 percent. In this way, a rancher could eliminate the need for a feedlot middleman and sell his fattened animals directly to slaughterhouses at a substantially greater price. To the extent that cattle may increasingly be kept on their Sandhills ranges rather than shipped as calves to cornbelt-centered cattle feedlots, this change could result in major shifts in regional economics and associated commodity-flow patterns.

Nebraska currently ranks second among all states in the total number of cattle and calves (5.9 million) and in annual cash receipts derived from cattle and calf production ($4.6 billion). It also ranks second in the number of cattle being maintained in feedlots (2.2 million). Because feedlots are major sources of nitrate pollution of surface water and groundwater supplies in Nebraska, there would seem to be potential ecological benefits in moving cattle away from such points of concentration. But center-pivots in the Sandhills pose serious ecological risks that often were not carefully considered by the developers, and the international market for corn is unpredictable at best. In 1990, when the total amount of Nebraska land under irrigation was about 8 million acres (3.2 million hectares), there were nearly 75,000 registered irrigation wells in the state. Corn prices throughout the 1970–90 period peaked at $2.92 per bushel during 1974 and again at $3.13 in 1983, reflecting international grain demands. Since then, the annual average price for corn has ranged from $1.52 to $2.58 per bushel and in recent years has averaged only slightly more than the typical farmer's production costs. Additionally, the values for irrigated land in Nebraska plummeted about 50 percent during the late 1980s, as government subsidy programs declined and foreign grain sales diminished.

The combination of reduced international corn sales and declining land values, occurring during a period of record high interest rates, high fuel costs, and increasing equipment costs, proved a recipe for financial disaster for many large-scale Sandhills irrigators in the mid-1980s. A federal farm bill passed in 1985 introduced some provisions intended to discourage irrigation of highly erodable soils, and a year later federal tax changes also began to discourage the conversion of natural wetlands to farmlands. Changes in federal tax laws in 1986 also made speculative agricultural investments for tax credits unattractive, especially to those individuals or investment groups being taxed at the highest rates.

Abandoned center-pivots, axle-deep in sand, became a powerful visual symbol of the highly speculative Sandhills developments of the 1970s and early 1980s and of the dangers of considering short-term dollar return over possible long-term ecological effects on the land.

CHAPTER TWELVE
The Enduring Land
Viewpoints and Visions

Fig. 46. Adult ferruginous hawk with rattlesnake.

OUR VIEW OF THE WORLD is myopic. Few of us can see far beyond our own limited horizons, to say nothing of peering into the future with any confidence. Our lifetimes are too short and our attitudes too strongly shaped by personal experience to allow us to look at the world very objectively. But we must do the best we can, hoping that our proper choices outnumber the mistaken ones and that the wrong roads taken are not one-way tracks like some Sandhills roads, where turnarounds cannot be made without danger of becoming deeply embedded in sand.

The economy of Nebraska is strongly tied to agriculture, and agriculture is in turn inextricably linked to water. Luckily, this most valuable resource is also the state's most abundant one. Yet like many seemingly unbounded riches, it is often undervalued, frequently squandered, and all too easily polluted. The Ogallala aquifer is a two-way street: water taken up from it can just as easily pass back down from the surface into the groundwater. That groundwater has been increasingly laced with nitrates, atrazine, and countless pesticides, herbicides, fungicides, and other biocides that are spread over our national landscape at a rate of about 1.1 billion pounds (0.5 billion kilograms) annually, an amount equivalent to 4.6 pounds (more than 2 kilograms) per U.S. citizen. Atrazine ("Atrex"), alochlor ("Lasso"), propachlor ("Ramrod"), 2,4-D, and several others have been detected in Nebraska's water supplies. Most of the newer biocides have not yet been independently tested or approved by the Environmental Protection Agency (EPA) as to their potential human health risks, and some older ones have been exempted from such testing because their use preceded legislation requiring pesticide screening for possible adverse health effects.

In Nebraska, about 33 million pounds (15 million kilograms) of pesticides are used annually—or 21 pounds (9.5 kilograms) per Nebraska resident—which ranks it seventh nationally among all states in total usage, although it is only fifteenth in overall surface area. About 1.94 million tons (1.76 metric tons) of nitrogen were applied to Nebraska farmlands in 1992 (a near-record amount), and some 235,000 additional tons (213,000 metric tons) of manure-related nitrogen are generated in the state annually as a result of cattle and hog production. When these materials infiltrate our surface water, our groundwater, and—through endless television and radio commercials for pesticides and herbicides— even our airwaves, we are at least metaphorically surrounded by deep and dangerous water. Nitrate levels in drinking waters that are in excess of the established federal limit of 10 parts per million (ppm) may lead to lethal methemoglobinemia ("blue baby syndrome") in infants. Yet in some Nebraska wells that are situated near barnyards, nitrate concentrations above 50 ppm are quite common, and the municipal water supplies of several Nebraska cities have been condemned for exceeding federal limits. Efforts toward cleaning the most highly polluted water supplies in the state, using uncounted millions of dollars of EPA superfund and other federal moneys, have had virtually no success in improving these polluted groundwater situations. Curiously, the Nebraska Water Resource Center, in its recently published list of "ordinary risks" to humans, informs us that statistically, Americans have a ten-times-greater chance of dying from eating four tablespoons of peanut butter daily than from being exposed to all current drinking water pollutants. Perhaps fortunately, if not coincidentally, Nebraska's farm economy is not dependent upon peanuts.

Nebraska's agricultural interests have dominated the state's use of water. As of 1985, agricultural activities accounted for 97 percent of all the state's consumptive water use, and about three-quarters of its total water use. Nebraska not only leads all the Great Plains states in land area currently under irrigation (nearly half the total) but during 1990 was second only to Texas in the amount of groundwater extracted: 4,790 million gallons (19.2 million cubic meters) per day (mgd). By comparison, the neighboring states of Wyoming and South Dakota then averaged less than 400 mgd each, or less than a tenth as much as Nebraska. Of the groundwater pumped in Nebraska in 1990, 4,360 mgd served to irrigate about 8 million acres (3.24 million hectares) with the remainder mainly used as drinking water by municipalities and individual well owners. The Ogallala aquifer provides 92 percent of the groundwater now available in Nebraska, and 86 percent of Nebraska's population in 1990 was dependent upon groundwater for their municipal water systems and individual wells. Most of these people were served by municipal systems, but about 20 percent of Nebraska's population received its drinking water from individual wells (235 mgd extracted for municipal water supplies, as compared with 47 mgd obtained from individual wells). Nebraska's current water laws fail to recognize the simple fact that surface water sources and groundwater supplies are interrelated; the state attempts to regulate only water obtained from rivers and other surface sources.

The depth and richness of the Ogallala aquifer in Nebraska is almost incredible; an estimated two-thirds of all the water in the entire aquifer lies under Nebraska, making it one of the richest known sources of groundwater in the world. It has been estimated that the total amount of water stored underground in the state could cover a flat surface the area of Nebraska to a depth of 34 feet (over 10 meters). And the aquifer is more than 400–600 feet (120–180 meters) thick below much of the Sandhills region with perhaps 20 percent of this volume being actual water (the rest comprising sand and gravel). As a result, the massive water-extraction rates have had only very slight effects on groundwater levels over most of the state. Although some significant reductions in the water table have occurred south of the South Platte and Platte Rivers, some corresponding rises have also occurred there as a result of local impoundments and irrigation activities (see figure 34).

Nevertheless, the state has paid a high price for its reckless exploitation of and virtual lack of control over groundwater usage. Because of the high water use, plus the closeness to the surface and the permeability of the Ogallala aquifer, Nebraska is particularly vulnerable. Nationally speaking, EPA studies indicate that about a third of all U.S. counties have both high pesticide use and also high vulnerability to groundwater pollution. The national survey found that about 10 percent of community wells and 4 percent of rural domestic wells contained residues of at least one pesticide, the most common being atrazine, DBCP (a nematocide), and DCPA metabolites. More than half the wells tested nationally also had detectable nitrate levels, but only 2 percent exceeded the federally accepted maximum (10 ppm) considered allowable for human consumption. By comparison, studies of 5,286 Nebraska wells during the middle and late 1980s, reported by M. E. Exner and R. F. Spalding, found that about 20 percent exceeded the federal safety standards for drinking water. The *average* annual concentrations of nitrates in the groundwaters of York, Polk, and Hamilton Counties in the Platte River Valley are now at least 5 ppm or more than half of the accepted limit. Interestingly, no more than a handful of wells in the Sandhills have thus far exceeded the nitrate safety limits, showing that even such extremely fragile landscapes *can* be managed safely and cared for responsibly (figure 47).

State and municipal agencies are slowly starting to deal with this enormous problem, mostly because of federally mandated require-

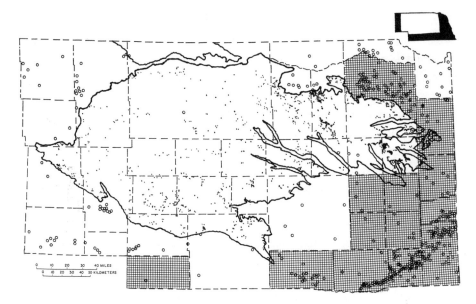

Fig. 47. Locations of wells having high nitrate levels (over 10 ppm) during the late 1980s (open circles), distribution of center-pivots within the Sandhills as of 1984 (dots), and major corn-growing areas (cross-hatched).

ments. The state legislature and its primary agency of environmental monitoring, the Department of Environmental Quality, have been notably lax in providing leadership in such important environmental issues as controlling both surface water and groundwater pollution. In fact, when the agency was formed in 1971 and called the "Environmental Control Council," 11 of its 16 initial members represented those interests in the state most responsible for causing pollution! Nebraska was the last state in the Union to pass enabling legislation (in 1993) bringing it into compliance with the 1947 Federal Insecticide, Pesticide, Fungicide, and Rodenticide Act (FIFRA), which requires the adequate training and certification of all biocide applicators and overseeing the enforcement of federal controls on pesticide use. Currently, about 45,200 certified pesticide applicators are present in Nebraska, of whom 85 percent are private agricultural producers and the remaining 15 percent are commercial applicators. As finally adopted, the legislation provides for Nebraska's pesticides to be regulated by its Department of Agriculture. A similar questionable allocation of state responsibilities now occurs in

Purviews and Prospects

Louisiana, where the Department of Agriculture and Forestry (DAF) oversees the state enforcement of federal pesticide regulations. Pesticide poisoning of an estimated 1 million fish and uncounted other aquatic life occurred in 1991, and yet following this massive environmental disaster, no fines have been collected by the Louisiana DAF. As a result of agriculturally biased attitudes, pesticide residue levels in Nebraska water supplies are likely to remain among the highest in the country.

Nebraska's Department of Environmental Quality is legally charged with the responsibility of investigating areas where political subdivisions (Groundwater Management Areas) have identified serious groundwater problems and must approve and oversee any plans for remedying them. The relevant Natural Resource District must then address identified water-quality problems within its jurisdiction, providing specific actions to control and perhaps ameliorate these problems. Groundwater Management Areas have been established by several of the Natural Resource Districts into which Nebraska is divided, including high-nitrate regions in the central Platte Valley, the "tri-basin" area directly to the south of the central Platte Valley in Gosper, Phelps, and Kearney Counties, and the Upper Elkhorn District in Rock and Holt Counties. Most of these problem areas are in the primary corn-growing regions of Nebraska, but relatively speaking, larger feedlot operations are the most serious sources of nitrate pollution in the state. Nitrogen in cattle wastes is rapidly converted by bacterial action to nitrates, which means that every feed-lot operation should be required to have fully effective sewage-control systems.

Closer to the Sandhills, serious nitrate contamination problems currently exist in Boyd County and the northern half of Brown County. It has been estimated that nearly all of the roughly 20 million acre-feet (24.7 billion cubic meters) of precipitation annually reaching the Sandhills surface actually enters the soil. Of this total, from 25 to 50 percent may penetrate deep enough to become part of the groundwater (the rest returning to the atmosphere by surface evaporation or plant transpiration). This means that through precipitation effects or excessive irrigation, nitrates and other soil-level pollutants are readily leached downward into substrate zones where they may contaminate the groundwater. Infiltration rates of up to ten feet (three meters) per day have been estimated in the Sandhills, meaning that there may be a nearly immediate transfer of some surface water to the water table, although rates of downward leaching of nitrates in the Sandhills have been estimated at only about five to seven feet (two meters or less) per year. In some of the shallow eastern Sandhills wells of Garfield and Wheeler Counties, higher than normal background amounts of nitrates, sulfates, chlorides, and dissolved solids have been found. These increased concentrations, as well as measurable amounts of the herbicide atrazine, have been attributed to the surface application of agricultural chemicals. For such areas as these the nitrate and pesticide groundwater pollution problems may soon become so severe that solutions will be beyond our control. These problems involve identifying and assigning possible legal responsibilities to the contaminators, understanding the ecological complexities of the damages that have already occurred, and attempting to rectify situations that are rapidly and literally becoming beyond our reach.

Related to the water-dependency problems associated with the center-pivot technology in the Sandhills is the equally great problem of soil and sand erosion resulting from the dune-leveling activities that are often performed during the installation of these systems. As reported by S. F. Hoesel, some lands in Brown County previously classified by the U.S. Department of Agriculture as unsuitable for irrigation have nevertheless been developed for center-pivot agriculture. Wind-erosion effects during the 40–60 hours that a center-pivot sprinkler requires to complete a revolution and rewet the dry and sandy soil can pose serious problems. Blowing sand might easily clog and bog down the entire center-pivot system, not only rendering it ineffective but also exposing the land to further erosive effects.

Sandhills ranchers have traditionally handled their lands with all the loving care that anybody could reasonably expect of good stewards. They and their ancestors have worked very hard for their properties; few groups of Nebraskans are more aware of their land-based heritage than are the people of the Sandhills. During the late 1980s, when some short-term financial interests tried to promote the sale of Nebraska water to out-of-state buyers whose water supplies were drying up, it was the Sandhills residents who banded together in the Preserve Our Water Resources Association to fight and eventually deflect this move. The Center for Rural Affairs, headquartered in Walthill, Thurston County, has since its formation in 1973 similarly voiced concern for the long-term effects of irrigation on the economy and ecology of the Sandhills. The legislative banning of farm corporations and limited partnerships for land ownership in Nebraska, which typically have little concern for the long-term ecological effects of their actions and deal only with short-term profits, represented a major victory for this group. Additionally, its members have worked to reduce government subsidies that provide financial rewards for converting native grasslands to irrigated croplands and have promoted government programs that help protect ground-water quality.

Ecologists and Sandhillers tend to operate as the opposite ends of an activity spectrum that ranges from engaging in highly theoretical mind games such as ecological modeling with computers, at one extreme, to being largely preoccupied with entirely practical, often land- or machinery-related activities at the other. Ecologists may, for example, be very good at calculating how much methane or nitrate is produced annually by a herd of cattle but are often very poor at fixing clogged drains or repairing fences. Sandhillers are necessarily good at repairing fences and fixing drains but are unlikely to spend much time thinking about the amount or effect of their cattle's methane production. We need both kinds of people, but, all things considered, Sandhillers are generally better company than ecologists. Certainly, if broken down on a back-country Sandhills road, there is no question about which sort of person I would want to see approaching in an old pickup over the dunes on the far horizon.

Appendices

Time Scale of Cenozoic Events in North America, with Special Reference to Nebraska and the Sandhills

Period	Epoch	Years before Present	Period or Stratum	Nebraska Examples
			Continental Glaciation	*Associated Deposits*
Quaternary	Holocene	0–12,000	Postglacial	dune shaping (Sandhills)
	Pleistocene	12–20,000	Post-Wisconsinian	dune building (Sandhills)
		20–150,000	Wisconsinian	alluvium, loess
			3d Interglacial	(eastern Nebraska)
		400–550,000	Illinoian	
			2d Interglacial	
		1–1.4 million	Kansan	sand to coarse gravel
			1st Interglacial	alluvium (eastern Nebraska and Platte Valley)
		1.7–2 million	Nebraskan	
			Geologic Stratum	*Associated Localities*
Tertiary	Pliocene	2 million	Broadwater Formation	poorly represented
		5 million		in Nebraska
		6 million		
	Miocene		Ogallala Group	
			Ash Hollow Formation	Poison Ivy Quarry
			Valentine Formation	Niobrara Canyon (north slope)
		18 million		
		24 million	Arikaree Group	
	Oligocene		Rosebud Formation	Niobrara (channel)
		28 million		
			White River Group	
			Brule Formation	Scotts Bluff (face)
		35 million		
		37 million		
	Eocene			not represented
		58 million		in Nebraska
	Paleocene			
		65 million		

Source: Adapted from Johnsgard 1984.

Ecological Checklists of Sandhills Vertebrates and Vascular Plants

Zoogeographic affinities for vertebrates (in parentheses after the name): E = eastern, G = grassland, N = northern, P = pandemic, S = southern, W = western.

Relative abundance for vertebrates: A = abundant, C = common, U = uncommon, L = local, O = occasional, R = rare.

MAMMALS

For this table only, E = eastern *or* northeastern, S = southern *or* neotropical, W = western *or* southwestern. Habitat preferences for prairie species: G = grasslands generally, M = moist meadows, S = Sandhills prairies, SG = shortgrass prairies, TG = tallgrass prairies. For marginal grassland mammals, X = more characteristic Great Plains habitats, x = its less typical habitats.

	Prairie	Shrub Steppe	Hardwood Forest	Pine Forest	Wetlands
Marsupials					
Opossum (s)	o (G)		X		
Insectivores					
Short-tailed shrew (E)	L (G)	x		X	
Masked shrew (N)	L (M)	x		X	
Least shrew (E)	L (G)	x		x	
Eastern mole (E)	L (M)	x			
Bats					
Keen's bat (E)			x	x	
Small-footed bat (P)			x	x	
Silver-haired bat (P)			x	x	x
Big brown bat (P)			x	x	x
Red bat (E)			x	x	x
Hoary bat (P)			x	x	x

	Prairie	Shrub Steppe	Hardwood Forest	Pine Forest	Wetlands
Rabbits					
Eastern cottontail (E)	c (G)	x	X		x
Desert cottontail (w)	l (G)	X			
Black-tailed jackrabbit (w)	u (sG)	X		x	
White-tailed jackrabbit (G)	u (G)	X		x	
Rodents					
Plains pocket gopher (G)	c (G)	x		X	
Plains pocket mouse (w)	A (s)	x			
Silky pocket mouse (w)	l (sG)	x			
Hispid pocket mouse (G)	A (G)	X			
Ord's kangaroo rat (w)	A (s)	X			
Beaver (P)			x		X
Western harvest mouse (w)	c (G)	X	X		
Plains harvest mouse (G)	u (G)	x			
Deer mouse (P)	c (G)	X	X	X	X
White-footed mouse (P)	l (G)	X	X		X
Northern grasshopper mouse (G)	u (s)	x			
Prairie vole (G)	c (TG)				
Meadow vole (N)	u (M)				X
Spotted ground squirrel (w)	u (s)	x			
Franklin's ground squirrel (G)	l (TG)				
Thirteen-lined ground squirrel (G)	A (G)	x			
Black-tailed prairie dog (G)	u (sG)	X		x	
Fox squirrel (E)			X		x
Muskrat (P)					X
Meadow jumping mouse (N)	u (M)		x		X
Porcupine (P)			x	X	
Carnivores					
Coyote (P)	c (G)	X	x	x	x
Red fox (P)	u (G)		X	X	x
Raccoon (P)	c (G)		X	x	X
Long-tailed weasel (P)	u (G)	x	x	x	X
Least weasel (N)	l (G)		x		X
Mink (P)			x		X
Badger (P)	u (G)	x			
Eastern spotted skunk (s)	l (G)	x	x		
Striped skunk (P)	c (G)	x	x	x	x
Bobcat (P)	u (G)	x	x	x	

	Prairie	Shrub Steppe	Hardwood Forest	Pine Forest	Wetlands
Ungulates					
Pronghorn (G)	O (SG)	x			
Elk (P)	R (SG)		x	x	
Bison (P)	R (SG)	x			
Mule deer (P)	C (G)	X	x	X	
White-tailed deer (P)	C (G)		X		X

Sources: Adapted largely from Freeman 1989b, excluding extirpated, very rare, and introduced species. General zoogeographic affinities after Jones, Armstrong, and Choate 1985.

BIRDS

G = species endemic to the Great Plains grasslands.

Grassland and/or Widespread Species	Woodland Species	Wetland Species

Breeding and Possibly Breeding(*) Permanent Residents

Grassland and/or Widespread Species	Woodland Species
Ferruginous hawk*	Red-tailed hawk (P)
Golden eagle*	American kestrel (P)
Prairie falcon*	Wild turkey (P)
Greater prairie-chicken (G)	Northern bobwhite (E)
Sharp-tailed grouse	Eastern screech-owl (E)
Mourning dove	Great horned owl (P)
Common barn-owl	Barred owl (E)*
Short-eared owl	Long-eared owl (P)
Eastern meadowlark	Red-bellied woodpecker (E)
Western meadowlark	Downy woodpecker (P)
	Hairy woodpecker (P)
	Northern flicker (P)
	Blue jay (E)
	Black-billed magpie (W)
	American crow (P)
	Black-capped chickadee (E)
	Red-breasted nuthatch (W)
	White-breasted nuthatch (P)
	Brown creeper (N)
	Northern cardinal (E)
	House finch (W)*

Grassland and/or Widespread Species	Woodland Species	Wetland Species

Breeding and Possibly Breeding(*) Migrants

Grassland and/or Widespread Species	Woodland Species	Wetland Species
Turkey vulture	Sharp-shinned hawk (P)	Pied-billed grebe
Northern harrier	Cooper's hawk (P)*	Western grebe
Swainson's hawk	American woodcock (E)*	Double-crested cormorant
Upland sandpiper (G)	Black-billed cuckoo (E)	American bittern
Long-billed curlew (G)	Yellow-billed cuckoo (E)	Least bittern*
Burrowing owl	Whip-poor-will (E)*	Great blue heron
Common nighthawk	Chimney swift (E)	Black-crowned night-heron
Common poorwill	Red-headed woodpecker (E)	White-faced ibis
Horned lark	Western wood-pewee (w)	Trumpeter swan
Northern rough-winged swallow	Eastern wood-pewee (E)	Canada goose
	Willow flycatcher (P)*	Wood duck
Bank swallow	Eastern phoebe (E)	Green-winged teal
Cliff swallow	Say's phoebe (w)	Mallard
Barn swallow	Great crested flycatcher (E)	Northern pintail
Loggerhead shrike	Cassin's kingbird (w)*	Blue-winged teal
Dickcissel (G)	Western kingbird (w)	Cinnamon teal
Field sparrow	Eastern kingbird (E)	Northern shoveler
Vesper sparrow	Purple martin (P)	Gadwall
Lark sparrow	Tree swallow (N)	American wigeon
Lark bunting (G)	House wren (E)	Canvasback
Savannah sparrow*	Eastern bluebird (E)	Redhead
Grasshopper sparrow	Mountain bluebird (w)*	Ring-necked duck*
Clay-colored sparrow (G)	Wood thrush (E)*	Lesser scaup
Chestnut-collared longspur (G)	American robin (P)	Ruddy duck
Bobolink	Gray catbird (E)	Virginia rail
Brewer's blackbird	Northern mockingbird (E)	Sora
	Brown thrasher (E)	Yellow rail*
	Cedar waxwing (P)	Common moorhen*
	Bell's vireo (E)	American coot
	Warbling vireo (P)	Piping plover
	Red-eyed vireo (E)	Killdeer
	Black-and-white warbler (E)*	Black-necked stilt
	Yellow warbler (P)	American avocet
	American redstart (E)	Willet
	Ovenbird (E)	Spotted sandpiper
	Yellow-breasted chat (P)	Common snipe
	Scarlet tanager (E)	Wilson's phalarope
	Rose-breasted grosbeak (E)	Forster's tern

Grassland and/or Widespread Species	Woodland Species	Wetland Species
	Black-headed grosbeak (w)	Least tern
	Blue grosbeak (s)	Black tern
	Lazuli bunting (w)*	Belted kingfisher
	Indigo bunting (e)	Sedge wren
	Rufous-sided towhee (p)	Marsh wren
	Chipping sparrow (p)	Common yellowthroat
	Song sparrow (p)	Swamp sparrow
	Common grackle (e)	Red-winged blackbird
	Brown-headed cowbird (p)	Yellow-headed blackbird
	Orchard oriole (e)	
	Northern oriole (p)	
	Pine siskin (n)	
	American goldfinch (p)	

Nonbreeding Migrants and Winter Visitors

Osprey	Northern goshawk (n)	Common loon
Bald eagle	Broad-winged hawk (e)	Horned grebe
Rough-legged hawk	Merlin (n)	Red-necked grebe
Peregrine	Ruby-throated hummingbird (e)	American white pelican
Gyrfalcon		Great egret
Snowy owl	Olive-sided flycatcher (n)	Snowy egret
Northern shrike	Yellow-bellied flycatcher (n)	Little blue heron
Baird's sparrow (g)	Alder flycatcher (n)	Cattle egret
LeConte's sparrow (g)	Least flycatcher (e)	Green-backed heron
Sharp-tailed sparrow	Western flycatcher (w)	Tundra swan
McCown's longspur (g)	Violet-green swallow (w)	Greater white-fronted goose
Lapland longspur	Winter wren (n)	Snow goose
Snow bunting	Golden-crowned kinglet (n)	Ross' goose
	Ruby-crowned kinglet (n)	Greater scaup
	Townsend's solitaire (w)	Common goldeneye
	Veery (n)	Bufflehead
	Gray-cheeked thrush (n)	Hooded merganser
	Swainson's thrush (n)	Common merganser
	Hermit thrush (n)	Sandhill crane
	Solitary vireo (n)	Whooping crane
	Philadelphia vireo (n)	Black-bellied plover
	Tennessee warbler (n)	Lesser golden-plover
	Orange-crowned warbler (w)	Semipalmated plover
	Nashville warbler (n)	Greater yellowlegs
	Magnolia warbler (n)	Lesser yellowlegs

Grassland and/or Widespread Species	Woodland Species	Wetland Species
	Yellow-rumped warbler (N)	Solitary sandpiper
	Black-throated green warbler (E)	Hudsonian godwit
		Marbled godwit
	Blackburnian warbler (N)	Red knot
	Palm warbler (N)	Sanderling
	Bay-breasted warbler (N)	Semipalmated sandpiper
	Blackpoll warbler (N)	Western sandpiper
	Cerulean warbler (E)	Least sandpiper
	Northern waterthrush (N)	White-rumped sandpiper
	Connecticut warbler (N)	Baird's sandpiper
	Mourning warbler (E)	Pectoral sandpiper
	McGillivray's warbler (w)	Dunlin
	Wilson's warbler (P)	Stilt sandpiper
	Canada warbler (N)	Short-billed dowitcher
	Western tanager (w)	Long-billed dowitcher
	American tree sparrow (N)	Red-necked phalarope
	Fox sparrow (N)	Franklin's gull (G)
	Lincoln's sparrow (N)	Bonaparte's gull
	White-throated sparrow (N)	Ring-billed gull
	White-crowned sparrow (N)	California gull
	Harris' sparrow (N)	Herring gull
	Dark-eyed junco (N)	Caspian tern
	Rusty blackbird (N)	Common tern
	Purple finch (N)	Water pipit
	Red crossbill (N)	
	Common redpoll (N)	
	Evening grosbeak (N)	

Sources: Based on information of Johnsgard 1986, Ducey 1988, and Labedz 1989, excluding all accidental and introduced species.

AMPHIBIANS AND REPTILES

Primary habitat preferences of grassland species (sequentially organized according to relative abundance): s = sandy, v = varied.

	Dryland Habitats	Temporary Wetlands	Permanent Wetlands	Brooks and Rivers
Grasslands Species				
Ornate box turtle (s)	A (s)			
Lesser earless lizard (w)	A (s)			
Northern prairie lizard (w)	A (v)			
Bull snake (G)	A (v)			
Six-lined racerunner (s)	C (s)			
Blue racer (P)	C (v)			
Western hog-nosed snake (w)	C (s)			
Plains gartersnake (P)	C (v)			
Many-lined skink (w)	U (s)			
Milk snake (w)	O (v)			
Prairie rattlesnake (w)	O (v)			
Semiaquatic and Aquatic Species				
Rocky Mountain toad (P)	A	A	A	
Plains spadefoot toad (w)	C	C		
Great Plains toad (G)		O		
Tiger salamander (P)	O	C	C	
Western striped chorus frog (N)		C		
Painted turtle (P)		O	C	
Bull frog (E)			L	
Snapping turtle (E)			C	
Yellow mud turtle (s)			L	
Blanding's turtle (N)			L	
Common gartersnake (w)			C	
Common watersnake (E)			L	L
Northern leopard frog (N)			L	L
Northern cricket frog (E)				L
Spiny softshell turtle (s, E)				L

Sources: Adapted from Lynch 1985 and esp. Freeman 1989a.

FISHES

For this table only, G = Great Plains or prairie. Where two or more habitat types are shown for a single species, the most typical Nebraska habitat is indicated by an upper-case X.

	Headwaters	Brooks	Larger Rivers	Lakes
Blacknose shiner (N)	X			
Northern redbelly dace (N)	X			
Finescale dace (N)	X	x		
Creek chub (E)	X	x		
Pearl dace (N)	X	x		
Brook stickleback (N)	x	X		
Plains topminnow (G)	x	X		
Iowa darter (N)	x	X		
Bigmouth shiner (N)		x		
Sand shiner (E, G)		x		
Longnose dace (w)		x		
Longnose sucker (N)		x		
Brassy minnow (N)		x		
Plains killifish (G)		x		
Suckermouth minnow (E)		x	x	
Central stoneroller (E, G)		X	x	
Goldeye (P)		x	X	
Plains minnow (G)		x	X	
Speckled chub (w)		x	X	
Flathead chub (w)		x	X	
Silver chub (P)		x	X	
Emerald shiner (P)		x	X	
River shiner (P)		x	X	
Stonecat (P)		x	x	
Tadpole madtom (E)		x	x	x
Gizzard shad (P)		x	x	x
Red shiner (G)		X	x	x
Fathead minnow (G)		X	x	x
Golden shiner (E)		x	x	X
River carpsucker (G)		x	X	x
Quillback (P)		x	X	x
White sucker (N)		X	x	x
Bigmouth buffalo (P)		x	X	x
Shorthead redhorse (P)		x	X	x
Black bullhead (G)		X	x	x
Yellow bullhead (E)		X	x	x

	Headwaters	Brooks	Larger Rivers	Lakes
Channel catfish (P)		x	X	x
Flathead catfish (P)		x	X	x
Grass pickerel (E)		x	X	x
Northern pike (P)		x	x	X
White bass (P)		x	x	X
Walleye (P)		x	x	X
Rock bass (E)		x	x	X
Green sunfish (G)		x	x	X
Orange-spotted sunfish (G)		x	x	X
Bluegill (E)		x	x	X
White crappie (E)		x	x	X
Black crappie (E)		x	x	X
Yellow perch (N)		x	x	X
Sauger (P)		x	x	X
Freshwater drum (P)		x	x	X

Source: Adapted with modifications from Hrabik 1989, excluding introduced, extirpated, and rare species.

VASCULAR PLANTS

Grasses are parenthetically identified as cool-season (C) or warm-season (W) species. Ecological affinities of primarily upland grassland species: BG = bunchgrass community or stable sand ridge and slope sites; BO = blowouts, eroding sand ridges, and barren sand draws; NT = needle-and-thread or valley sites; SG = shortgrass (grama and buffalo grass) community; SM = sand muhly community or unstable slope and sandy disturbance sites; TA = three-awn community or Sandhills–shortgrass prairie transition sites; WM = wheatgrass meadow community or dry meadow and swale sites. Ecological affiliations for primarily lowland and aquatic species: A = aquatic, E = emergent, LM = lowland meadow, S = shoreline. Primary ecological affiliations of tree species: F = floodplains, NS = north-facing slopes, and RU = rocky uplands.

	Grasslands	Woodlands	Wetlands
Upland Grasses and Sedges			
Agropyron spp. (wheatgrasses, C)	WM		
Andropogon hallii (sand bluestem, W)	BG, SM, NT		
Andropogon scoparius (little bluestem, W)	BG, NT		
Spartina spp. (cordgrasses, W)	WM		
Aristida spp. (three-awn grasses)	TA		
Buchloë dactyloides (buffalo grass, W)	TA, SG		

	Grasslands	Woodlands	Wetlands
Bouteloua gracilis (blue grama, w)	TA, SG		
Bouteloua hirsuta (hairy grama, w)	BG, SM, TA		
Calamogrostris stricta (northern reedgrass)	WM		
Calamovilfa longifolia (prairie sandreed, w)	BG, SM, BO		
Carex spp. (sedges)	BG		S
Cyperus spp. (umbrella grasses)	BG		S
Eragrostis trichodes (sand lovegrass, w)	BG, BO		
Koeleria cristata (Junegrass, C)	NT		
Muhlenbergia pungens (sand muhly, w)	SM, BO		
Oryzopsis hymenoides (ricegrass, C)	BG, BO		
Panicum virgatum (switchgrass, w)	BG, M		
Redfieldia flexuosa (blowout grass, w)	BO		
Sorghastrum avenaceum (Indian grass, w)	BG, BO		
Sporobolus crypandrus (sand dropseed, w)	BG		
Stipa comata (needle-and-thread, C)	BG, SM, NT, TA		
Stipa spartea (porcupine grass, C)	TA		
Stipa spp. (needlegrasses)	WM		

Upland Forbs and Small Shrubs

	Grasslands	Woodlands	Wetlands
Ambrosia psilostachya (western ragweed)	BO, SM		
Amorpha canescens (leadplant)	BG		
Anemone caroliniana (Carolina anemone)	BG		
Antennaria parviflora (Rocky Mountain pussytoes)	BG		
Argemone polyanthemos (prickly poppy)	SM		
Artemisia ludoviciana (white sage)	BG		
Artemisia spp. (sages)	SM		
Asclepias arenaria (sand milkweed)	BG		
Asclepias latifolia (broadleaf milkweed)	BG		
Asclepias viridiflora (green milkweed)	BG		
Calylophus serrulata (plains yellow primrose)	BO		
Ceanothus herbaceous (New Jersey tea)	BG		
Chenopodium spp. (goosefoots)	SM		
Chrysopsis villosa (golden aster)	BG, SM		
Cirsium canescens (Platte thistle)	SM		
Cleome serrulata (Rocky Mountain beeplant)	SM, WM		
Collomia linearis (collomia)	SM		
Commelina virginica (dayflower)	BG		
Coryphantha vivipara (pincushion cactus)	BG		
Croton texensis (croton)	BO		
Cryptantha celosioides (miner's candle)	BG, BO		
Cyclomoma atriplicifolia (winged pigweed)	BO		
Delphinium virescens (prairie larkspur)	BG, WM		

	Grasslands	Woodlands	Wetlands
Eriogonum annuum (annual eriogonum)	BG		
Erysimum asperum (western wallflower)	BG		
Euphorbia spp. (spurges)	BG		
Froelichia floridana (field snake-cotton)	BO, SM		
Haplopappus spinulosus (cutleaf ironplant)	BG		
Helianthus rigidus (stiff sunflower)	BO		
Helianthus petiolaris (plains sunflower)	BG		
Ipomoea leptophylla (bush morning-glory)	BG, SM		
Ipomopsis longiflora (gilia)	BG		
Lathyrus polymorphus (showy vetchling)	BO		
Lesquerella argentea (bladderpod)	BG		
Liatrus punctata (dotted gayfeather)	BG		
Lithospermum carolinensis (hairy puccoon)	BG		
Lithospermum incisum (narrow-leaved puccoon)	BG, SM		
Mentzelia spp. (mentzelias)	SM		
Oenothera nuttallii (white-stemmed evening primrose)	SM, BO		
Oenothera rhombipetala (fourpoint evening primrose)	BG		
Opuntia spp. (prickly-pear cactus)	BG, SM		
Oxytropis lambertii (woolly locoweed)	BO, BG		
Penstemon angustifolius (narrow beardtongue)	BG		
Penstemon grandiflorus (large beardtongue)	BG		
Penstemon haydenii (Hayden's penstemon)	BO		
Petalstemon purpureum (purple prairie-clover)	BG, BO		
Petalostemon villosum (silky prairie-clover)	BG, BO		
Phlox andicola (creeping phlox)	BO, BG, WM		
Polanisia dodecandra trachysperma (clammy-weed)	BO		
Polanisia jamesii (cristatella)	BO		
Psoralea lanceolata (lemon scurfpea)	BG		
Psoralea tenuiflora (wild alfalfa)	NT		
Psoralea spp. (scurfpeas)	BG, BO		
Rumex venosus (sand dock/wild begonia)	BG, SM		
Ratibida columnifera (prairie coneflower)	WM		
Senecio plattensis (prairie ragwort)	WM		
Thelesperma megapotamicum (nippleweed)	BG		
Tradescantia occidentalis (prairie spiderwort)	BG, WM		
Verbena stricta (hoary vervain)	SM		

Lowland and Aquatic Grasses and Sedges

	Grasslands	Woodlands	Wetlands
Agrostis hyemalis (ticklegrass)	LM		S
Agrostis stolonifera (red-top, c)	LM, WM		
Beckmannia syzigachne (American sloughgrass)			S
Distichlis spicata (saltgrass, w)	LM		S

	Grasslands	Woodlands	Wetlands
Eleocharis spp. (spikerushes)	LM		S, E
Elymus canadensis (wild rye, C)	LM, WM		
Fimbristylis spp. (fimbristylis)			S
Glyceria spp. (mannagrass)			S, E
Hordeum jubatum (foxtail barley, C)	LM		S
Hemicarpha micrantha (common hemicarpha)			S
Juncus spp. (rushes)	LM		S, E
Leersia oryzoides (whitegrass)			S
Leptochloa fascicularis (bearded spangletop)			S
Phalaris arundinacea (reed-canary grass)	LM		S
Phragmites australis (common reed)			S, E
Spartina pectinata (prairie cordgrass, W)	LM, WM		
Sphenopholis obtusata (prairie wedgegrass)			S
Scirpus spp. (bulrushes)			S, E
Typha spp. (cattails)			S, E
Zizania aquatica (wild rice)			E

Lowland and Aquatic Forbs

	Grasslands	Woodlands	Wetlands
Alisma spp. (water plantains)			A
Asclepias incarnata (swamp milkweed)	LM		S
Azolla spp. (floating azollas)			A
Berula erecta (water parsnip)			S
Bidens spp. (beggar-ticks)			S
Bidens coronata (tickseed sunflower)			S
Campanula aparinoides (marsh bellflower)	LM		
Ceratophyllum demersum (coontail)			A
Cicuta spp. (water hemlocks)	LM		S
Dryopteris cristata (crested shieldfern)	LM		
Elodea spp. (waterweeds)			A
Galium spp. (bedstraws)	LM		
Helenium autumnale (sneezeweed)			S
Helianthus nuttalli (Nuttall's sunflower)			S
Lilium philadelphicum (western red lily)	LM		S
Lemna spp. (duckweeds)			A
Lobelia spicata (pale-spike lobelia)			S
Lobelia syphilitica (blue cardinal-flower)			S
Lycopus asper (rough bugleweed)			S
Lycopus americanus (American bugleweed)			S
Lysimachia thrysiflora (tufted loosestrife)	LM		
Lythrum dacotanum (loosestrife)			S
Mentha arvensis (field mint)	LM		
Mimulus spp. (monkeyflowers)	LM		

	Grasslands	Woodlands	Wetlands
Myriophyllum spp. (water milfoils)			A
Najas spp. (naiads)			A
Nasturtium officinale (watercress)			A
Nuphar luteum (cowlily)			A
Nymphaea spp. (waterlilies)			A
Onoclea sensibilis (sensitive fern)	LM		S
Polygonum spp. (smartweeds)			S, E
Potamogeton spp. (pondweeds)			E, A
Ranunculus cymbalaria (shore buttercup)	LM		S
Ranunculus spp. (water crowfoots)			A
Rumex maritimus (golden dock)			S
Ruppia maritima (wigeon grass)			A
Sagittaria spp. (arrowheads)			S, E
Scutellaria galericulata (marsh skullcap)	LM		
Scutellaria latiflora (blue skullcap)			S
Sparganium eurycarpum (bur-reed)			S
Spirodella polyrhiza (giant duckweed)			A
Stachys palustris (hedge nettle)			S
Teucrium canadense (American germander)	LM		
Thelypteris palustris (marsh fern)	LM		
Triglochin maritima (arrow grass)	LM		S
Utricularia spp. (bladderworts)			A
Verbena hastata (blue vervain)			S
Wolffia spp. (watermeal)			A
Zannichellia palustris (horned pondweed)			A

Larger Shrubs, Vines, and Small Trees

	Grasslands	Woodlands	Wetlands
Amelanchier alnifolia (serviceberry)		x	
Amorpha fruticosa (false indigo)	x	x	
Cornus amomum (pale dogwood)		x	
Cornus stolonifera (red-osier dogwood)			x
Prunus americana (American plum)	x	x	
Prunus besseyi (sand cherry)	x	x	
Prunus virginiana (chokecherry)	x	x	
Parthenocissus vitacea (Virginia creeper)		x	
Rhamnus lanceolata (buckthorn)		x	
Rhus aromatica (skunkbush sumac)		x	
Rhus glabra (smooth sumac)		x	
Rosa arkansana (Arkansas rose)	x		
Rosa woodsii (western wild rose)	x		
Salix amygdaloides (peachleaf willow)		x	x
Salix exigua (sandbar willow)	x		x
Salix humilis (prairie willow)	x		

	Grasslands	Woodlands	Wetlands
Sambucus canadensis (elderberry)		x	
Shepherdia argentea (buffaloberry)		x	
Symphoricarpos occidentalis (snowberry)	x		
Toxicodendron spp. (poison ivys)	x	x	
Vitis spp. (grapes)		x	
Yucca glauca (small soapweed)	x	x	
Zanthoxylum americanum (prickly ash)		x	

Deciduous Trees

	Grasslands	Woodlands	Wetlands
Acer negundo (box elder)		F	
Acer saccharinum (silver maple)		F	
Betula papyrifera (paper birch)		NS	
Celtis occidentalis (hackberry)		F	
Fraxinus pennsylvanica (green ash)		F	
Gleditsia triacanthos (honeylocust)		F	
Juglans nigra (black walnut)		F	
Quercus macrocarpa (bur oak)		NS	
Ostrya virginiana (ironwood)		F	
Populus deltoides (eastern cottonwood)		F	
Tilia americana (basswood)		F	
Ulmus americana (American elm)		F	

Coniferous Trees

	Grasslands	Woodlands	Wetlands
Juniperus virginana (eastern redcedar)		NV	
Juniperus scopulorum (Rocky Mountain juniper)		RU, NS	
Pinus ponderosa (ponderosa pine)		RU, NS	

Sources: Adapted from Pool 1914; Tolstead 1942; Keeler, Harrison, and Vescio 1980; Novacek 1989; and Kaul 1989, excluding most introduced and rarer indigenous species.

APPENDIX THREE
Distributional Checklist of Sandhills Birds

Wildlife refuges: v = Valentine, CL = Crescent Lake, FN = Fort Niobrara, L = Lacreek.

Common loon. Uncommon spring and fall migrant, early May to late May and late October to early November, but with summer occurrences of non-breeders on larger waters.

Pied-billed grebe. Common spring and fall migrant and local breeder (L, CL, V), early April to early November.

Horned grebe. Uncommon spring and fall migrant, mid-April to early May and early October to early November.

Red-necked grebe. Rare spring and fall migrant, with Nebraska records from mid-March to early May and late September to late October.

Eared grebe. Common spring and fall migrant and local breeder (L, CL, V), late April to mid-October.

Western grebe. Common spring and fall migrant and local breeder (L, CL, V), early May to early October. The recently described Clark's grebe, an apparently separate species closely resembling the western grebe, has been observed near the Sandhills in Keith County and no doubt also occasionally occurs on Sandhills lakes.

American white pelican. Common spring and fall migrant and local breeder (L), late April to late May and late September to mid-October; nonbreeders are often present on larger waters through the summer.

Double-crested cormorant. Common spring and fall migrant and local breeder (L, CL), in nonbreeding areas mid-April to early May and late September to late October.

American bittern. Common spring and fall migrant and local breeder (L, CL, V), early May to early October.

Least bittern. Uncommon to rare spring and fall migrant and probable local breeder, mid-May to mid-August.

Great blue heron. Common spring and fall migrant and local breeder, especially in woodlands near rivers and larger lakes (L, CL), early April to mid-October in breeding areas.

Great egret. Occasional to rare spring and fall migrant, mainly eastwardly, mid-April to early May and early August to mid-October.

Snowy egret. Occasional to rare spring and fall migrant, mid-April to early June and late July to early October.

Little blue heron. Occasional to rare spring and fall migrant, early April to early June and late July to late October.

Cattle egret. Rare but increasingly regular spring and fall migrant and summering nonbreeder or possible local breeder, early May to early September. A nesting pair was seen (v) in 1982, and more than 20 birds were present in 1984.

Green-backed heron. Common to occasional spring and fall migrant and probable local summer resident, but nesting so far reported only for Cherry County, late April to mid-September.

Black-crowned night-heron. Common spring and fall migrant and local breeder (L, CL, v), late April to early September.

White-faced ibis. Rare spring and fall migrant and very local breeder (CL, v) in recent years, mid-April to early October.

Tundra swan. Occasional to rare spring and fall migrant, mainly March–April and October–November.

Trumpeter swan. Extirpated historically as a breeder (before 1890) but now a rare spring and fall migrant and local breeder (L) as a result of restoration efforts (Ducey 1984), late March to early October and probably overwintering occasionally on open water.

Greater white-fronted goose. Common spring and fall migrant, mid-March to mid-April and late October to early November.

Snow goose. Common spring and fall migrant, early March to late April and early October to early December.

Ross' goose. Occasional to rare spring and fall migrant (often seen in company with snow geese and sometimes hybridizing with them), mid-March to mid-April and early November to late November.

Canada goose. Common spring and fall migrant (various races) and, as a result of restoration efforts of the race *maxima* (Gabig 1986), an increasingly widespread breeder (L, CL, v), late March to early December and larger races regularly overwintering on open water.

Wood duck. Uncommon to rare spring and fall migrant and local breeder (Sandhills counties breeding records for Garden, Keith, and Brown Counties), late March to late October.

Green-winged teal. Abundant to common spring and fall migrant and widespread breeder (L, CL, FN, v), mid-March to early November.

American black duck. Rare spring and fall migrant, spring records early March to late May, fall records August to late December.

Mallard. Abundant spring and fall migrant and widespread breeder (L, CL, FN, v), mid-March to late November.

Blue-winged teal. Abundant spring and fall migrant and widespread breeder (L, CL, FN, v), early April to mid-October.

Northern pintail. Abundant spring and fall migrant and widespread breeder (L, CL, FN, v), mid-March to late November.

Cinnamon teal. Uncommon spring and fall migrant and local breeder (CL), late April to mid-September. Recent regular during summer (up to four pairs seen in 1990 and 1991) and at least a periodic breeder at CL.

Northern shoveler. Common to abundant spring and fall migrant and widespread breeder (L, CL, FN, V), late March to early November.

Gadwall. Common to abundant spring and fall migrant and widespread breeder (L, CL, FN, V), late March to late November.

American wigeon. Common to abundant spring and fall migrant and uncommon to occasional breeder (L, CL, FN, V), late March to mid-November.

Canvasback. Common to uncommon spring and fall migrant and uncommon but widespread breeder (L, CL, FN, V), mid-March to mid-November.

Redhead. Common spring and fall migrant and widespread breeder (L, CL, FN, V), mid-March to early November.

Ring-necked duck. Common to uncommon spring and fall migrant and extirpated or accidental breeder (old records for CL, FN, V); pre-1920 breeding records for Brown, Cherry, Garden, and Morrill Counties), late March to late April and mid-October to late November.

Lesser scaup. Common to abundant spring and fall migrant and occasional local breeder (L, CL, FN), mid-March to mid-May and mid-October to late November.

Common goldeneye. Common to uncommon spring and fall migrant, early to late March and mid-November to mid-December.

Bufflehead. Common to uncommon spring and fall migrant, mid-March to mid-April and late October to late November.

Hooded merganser. Uncommon to occasional spring and fall migrant and previous rare breeder (Cherry County, pre-1920), late March to late April and early to late November; probable nonbreeders sometimes present during summer.

Common merganser. Common to uncommon spring and fall migrant, early March to early April and mid-November to mid-December; occasional to accidental nester (Cherry County pre-1920, Custer County in 1968).

Ruddy duck. Common to uncommon spring and fall migrant and local breeder (L, CL, V), early April to early November.

Turkey vulture. Uncommon to common spring and fall migrant and uncommon to occasional breeder (L, CL, FN, V; post-1960 breeding records for Keya Paha, Brown, Cherry, and Sheridan Counties), mid-April to late September.

Osprey. Uncommon to occasional spring and fall migrant, mid-April to early May and mid-September to early October.

Bald eagle. Uncommon spring and fall migrant and locally common overwintering migrant on open water (pre-1900 breeding evidence for Cherry County; recent accidental breeding attempts in Garden County), late November to mid-March.

Northern harrier. Uncommon spring and fall migrant and widespread but uncommon breeder (L, CL, FN, V), mid-March to early December.

Sharp-shinned hawk. Occasional to uncommon late fall and early spring or overwintering migrant, chiefly mid-September to mid-November and early January to mid-April, but with many December records. There is an early

1980s nesting record from the Niobrara Valley of Brown County and a 1981 nesting record for Sherman County.

Cooper's hawk. Occasional to uncommon late fall and early spring or overwintering migrant, mid-September to early November and early January to mid-April, but with many December records.

Northern goshawk. Occasional overwintering migrant, early January and mid-April to mid-May.

Broad-winged hawk. Occasional spring and fall migrant, mainly eastwardly, late April to mid-May and mid-September to early October.

Swainson's hawk. Common spring and fall migrant and widespread breeder (L, CL, FN, V), with breeding records for most Sandhills counties, especially westwardly, from mid-April to late September.

Red-tailed hawk. Common permanent resident and widespread breeder in wooded areas (L, FN, V), with breeding records for most Sandhills counties.

Ferruginous hawk. Occasional permanent resident, especially in the Panhandle's high plains; occasional to rare breeder in the western Sandhills (L, CL), with post-1960 breeding records for Cherry, Hooker, Garden, and Lincoln Counties.

Rough-legged hawk. Uncommon overwintering migrant, especially westwardly, early November to late March.

Golden eagle. Uncommon spring and fall migrant and local permanent resident at the western edge of the Sandhills, with post-1960 breeding records for Box Butte, Sheridan, Garden, Morrill, Keith, and Lincoln Counties.

American kestrel. Common permanent resident, becoming uncommon to occasional during winter, and local breeder (L, CL, FN, V), with scattered county breeding records for the Sandhills (Keya Paha, Brown, McPherson, Keith, and Lincoln Counties).

Merlin. Uncommon overwintering migrant, late October to mid-March but probably most common in December.

Peregrine falcon. Rare to occasional spring and fall migrant and overwintering migrant, late September to late March, with scattered records through the winter months.

Gyrfalcon. Extremely rare overwintering migrant, with state records late November to early March.

Prairie falcon. Occasional to rare permanent resident in the Panhandle and occasional apparent spring and fall or overwintering migrant in the Sandhills. There are post-1960 breeding records for Sheridan, Garden, and Morrill Counties and earlier breeding records for Cherry and Keith Counties.

Ring-necked pheasant. Introduced common to abundant permanent resident and widespread breeder (L, CL, FN, V).

Gray partridge. Introduced local permanent resident mainly in eastern and northern areas; has reportedly bred at V.

Greater prairie-chicken. Uncommon to occasional permanent resident especially in the eastern Sandhills, widespread breeder (L, CL, FN, V) with post-1960 records for Cherry, Thomas, Blaine, McPherson, Keith, and Lincoln Counties.

Sharp-tailed grouse. Uncommon to abundant permanent resident, especially westwardly, throughout the Sandhills. Widespread breeder (L, CL, FN, V) with post-1960 records for Cherry, Sheridan, Garden, Keith, Lincoln, Arthur, McPherson, Thomas, and Blaine Counties.

Wild turkey. Uncommon to occasional permanent resident, mainly in wooded river valleys, and a local breeder (FN); post-1960 summer breeding records for Keya Paha, Brown, Cherry, Sheridan, Keith, and Lincoln Counties.

Northern bobwhite. Uncommon to occasional permanent resident, mainly in the eastern Sandhills, and a fairly widespread breeder (L, CL, FN, V) with post-1960 records for Keya Paha, Cherry, Keith, and Lincoln Counties.

Yellow rail. Extremely rare spring and fall migrant in Nebraska late April to early June (no fall records), and possibly very rare summer resident in the northern Sandhills (L).

Virginia rail. Uncommon spring and fall migrant and widespread breeder (L, CL, FN, V), early May to mid-September.

Sora. Common spring and fall migrant and widespread breeder (L, CL, FN, V), early May to late September.

Common moorhen. Occasional spring and fall migrant, mid-May to late August and possible rare summer resident, especially eastwardly (CL and V; has reportedly bred at V).

American coot. Common to abundant spring and fall migrant and widespread breeder (L, CL, FN, V), post-1960 breeding records from Cherry, Sheridan, Morrill, Garden, and Lincoln Counties, late March to early November.

Sandhill crane. Abundant spring and fall migrant (mainly representing the race *canadensis*) and now extirpated but historic breeder (until 1883) in the Sandhills (the race *tabida*), early March to early April and early October to early November.

Whooping crane. Very rare spring and fall migrant and probable historic breeder in the Sandhills, April and October, averaging later in the spring and earlier in the fall than the sandhill crane.

Black-bellied plover. Uncommon to rare spring and fall migrant, early to late May and late August to early October.

Lesser golden-plover. Uncommon to occasional spring and fall migrant, late April to mid-May and late September to mid-October.

Semipalmated plover. Uncommon to occasional spring and fall migrant, late April to late May and mid-August to mid-September.

Piping plover. Occasional spring and fall migrant and probable local summer resident, early May to mid-August. Reported during summer L and V; breeding locally on alkaline wetlands of western Sandhills (Tim Cramer, pers. comm.). Post-1960s breeding records exist for Holt, Rock, Brown, Keith, and Lincoln Counties.

Killdeer. Common to abundant spring and fall migrant and breeder throughout, mid-March to late October.

Black-necked stilt. Rare spring and fall migrant and very rare breeder in the western Sandhills (has probably bred regularly at CL since 1985, and near Lakeside, in Sheridan County), late April to early August.

American avocet. Uncommon spring and fall migrant and widespread breeder (L, CL, FN, V), with post-1960 county breeding records for Cherry, Sheridan, Morrill, and Garden Counties, late April to early September.

Greater yellowlegs. Common spring and fall migrant, mid-April to early May and mid-August to early October.

Lesser yellowlegs. Common spring and fall migrant, mid-April to mid-May and mid-August to early October.

Solitary sandpiper. Uncommon to occasional spring and fall migrant, early to mid-May and early August to early September.

Willet. Uncommon to common spring and fall migrant and widespread breeder (L, CL, FN, V), with post-1960 breeding records for Cherry, Sheridan, and Garden Counties, late April to late August.

Spotted sandpiper. Common spring and fall migrant and widespread breeder (L, CL, FN, V), with post-1960 breeding records for Keya Paha and Greeley Counties, early May to early September.

Upland sandpiper. Uncommon spring and fall migrant and widespread breeder (L, CL, FN, V), with post-1960 breeding records for most western Sandhills counties, early May to late August.

Whimbrel. Extremely rare spring and fall migrant, with most Nebraska records for spring, mid-April to late May.

Long-billed curlew. Common spring and fall migrant and widespread breeder (L, CL, FN, V), with post-1960 breeding records for Brown, Cherry, Sheridan, Box Butte, Garden, Arthur, Keith, McPherson, Logan, and Lincoln Counties, from mid-April to mid-August. Proximity between moist meadows for foraging and grassy upland for nesting provides ideal nesting habitat in the Sandhills (Bicak 1977).

Hudsonian godwit. Rare spring migrant, mainly eastwardly, late April to mid-May.

Marbled godwit. Uncommon spring and fall migrant and very local breeder in extreme northern Sandhills (L) but no definite Nebraska breeding records, late April to early May and late July to early October.

Red knot. Extremely rare spring and fall migrant in the eastern Sandhills, early to mid-May and late August to October.

Sanderling. Occasional to rare spring and fall migrant, early to mid-May and mid-August to early October.

Semipalmated sandpiper. Common spring and fall migrant, late April to mid-May and early August to mid-September.

Western sandpiper. Rare spring and fall migrant, mainly westwardly, early to mid-May and mid-August to early September.

Least sandpiper. Common to occasional spring and fall migrant, early to mid-May and early August to mid-September.

White-rumped sandpiper. Common to occasional spring migrant, late April to mid-May and late July to early October.

Baird's sandpiper. Common to abundant spring and fall migrant, late April to mid-May and mid-August to early October.

Dunlin. Occasional to extremely rare spring migrant, mostly eastwardly, early to late May and September.

Stilt sandpiper. Uncommon to occasional spring and fall migrant, early to mid-May and mid-August to late September.

Short-billed dowitcher. Occasional to rare spring and fall migrant, mainly eastwardly, April and August–September.

Long-billed dowitcher. Common to uncommon spring and fall migrant, early to mid-May and early August to mid-October.

Common snipe. Common to uncommon spring and fall migrant and an apparently local breeder (L, CL), with post-1960 breeding records for Rock, Sheridan, and Garden Counties, mid-April to mid-November, sometimes overwintering.

American woodcock. Uncommon spring and fall migrant regarded by Labedz (1989) as a possible rare breeder, mid-March to early June and mid-September to mid-November.

Wilson's phalarope. Common to abundant spring and fall migrant and local breeder (L, CL) in the alkaline wetlands of the western Sandhills (Bomberger 1983), early May to early September.

Red-necked phalarope. Uncommon to rare spring and fall migrant, mainly westwardly, early to late May and early August to late September.

Franklin's gull. Common to abundant spring and fall migrant and very local breeder (occasional nester at L; 1966 and 1967 breeding records for CL), early April to mid-May and early September to mid-October, nonbreeders or immatures sometimes summering in Sandhills region.

Bonaparte's gull. Occasional to rare spring and fall migrant, mid-April to early May and October–November.

Ring-billed gull. Uncommon spring and fall migrant, mid-March to mid-May and mid-September to late November, immatures or nonbreeders sometimes remaining through summer; occasional breeding likely (since breeding occurs in eastern Wyoming) but still unproven.

California gull. Very rare spring and fall migrant, mainly in the western Sandhills (reported from Garden and Lincoln Counties), mid-March to early April and mid-July to late October. Because breeding occurs locally in eastern Colorado and east-central Wyoming, Nebraska breeding may eventually occur.

Herring gull. Uncommon spring and fall migrant, mid-March to late April and late October to late November, nonbreeders sometimes present during summer.

Caspian tern. Occasional spring and fall migrant, early to mid-May and September, but nonbreeders recorded from June onward.

Common tern. Rare to occasional spring and fall migrant, early to mid-May and mid-August to early September.

Forster's tern. Common spring and fall migrant and uncommon but apparently widespread breeder (L, CL, FN, V), with post-1960 breeding records for Cherry and Garden Counties, late April to mid-September.

Least tern. Uncommon to rare spring and fall migrant and possible breeder, mid-May to late September. Probably not breeding within the Sandhills proper, but numerous post-1960 breeding records exist for the Platte and Niobrara Valleys and for the Calamus Valley (Valley County).

Black tern. Common to abundant spring and fall migrant and widespread breeder (L, CL, FN, V), with post-1960 breeding records for Cherry, Garden, and McPherson Counties, mid-May to late September.

Rock dove. Introduced permanent resident around farms, ranches, and towns.

Mourning dove. Common to abundant spring and fall migrant and breeder (L, CL, FN, V), with post-1960 breeding records for nearly all Sandhills counties, where nesting usually occurs in trees (Klataske 1966) but sometimes on the ground in treeless habitats, late March to early November in some mild winters a year-around resident.

Black-billed cuckoo. Uncommon spring and fall migrant and uncommon but widespread breeder (CL, FN, V), with scattered post-1960 nesting records for Brown, Keith, and Lincoln Counties, late May to late August.

Yellow-billed cuckoo. Uncommon spring and fall migrant and local occasional to rare local breeder (L, V), with post-1960 breeding records for Cherry, Logan, Lincoln, and Keith Counties.

Common barn-owl. Uncommon permanent resident, especially southwardly, and local breeder (CL), with post-1960 breeding records for most Sandhills counties, where nesting often occurs in recesses or burrows of steep dune slopes; relatively common in the mixed grassland-cropland area at the southern edge of the Sandhills (Gubanyi et al. 1992).

Eastern screech-owl. Uncommon to rare permanent resident and fairly widespread breeder (L, CL, FN, V), with post-1960 breeding records for Cherry, Box Butte, and Lincoln Counties.

Great horned owl. Uncommon permanent resident and widespread breeder (L, CL, FN, V), with post-1960 breeding records for nearly all Sandhills counties.

Snowy owl. Occasional overwintering migrant, December to February.

Burrowing owl. Uncommon summer resident, mid-April to mid-September, and widespread breeder (L, CL, FN, V), with post-1960 breeding records for Brown, Sheridan, Garden, Keith, and Lincoln Counties.

Barred owl. Possible rare permanent resident in extreme east but probably not in the Sandhills proper.

Long-eared owl. Occasional permanent resident, probably limited to larger river valleys, with breeding reported in L and post-1960 breeding records for Cherry, Sheridan, Boone, and Lincoln Counties.

Short-eared owl. Common to occasional permanent resident and local refuge breeder (L, CL), with post-1960 breeding records for Garden and Lincoln Counties.

Northern saw-whet owl. Occasional to rare overwintering migrant, early November to late February; possibly a permanent resident at northwestern edge of the Sandhills in ponderosa pine forests and listed as a breeder at V, but no documented state breeding records exist.

Common nighthawk. Common to abundant spring and fall migrant and widespread breeder (L, CL, FN, V), with post-1960 breeding records for many Sandhills counties, late May to mid-September.

Common poorwill. Uncommon to rare spring and fall migrant and very local summer resident (FN) in rocky habitats at the edges of the Sandhills, early May to early September; only one post-1960 county breeding record for the Sandhills region, for the Niobrara Valley of Brown County, plus Panhandle records for Sheridan County.

Whip-poor-will. Occasional to rare spring and fall migrant and possible breeder at the eastern edge of the Sandhills, May to early September. A historic breeding record exists for Antelope County, and singing males were heard in Keya Paha, Brown, and Cherry Counties during 1982.

Chimney swift. Common to uncommon spring and fall migrant, mainly eastwardly, and probable local breeder in villages or cities of the Sandhills region, with post-1960 breeding records for Lincoln and Brown Counties, late April to early October.

Ruby-throated hummingbird. Uncommon to occasional spring and fall migrant, mainly eastwardly, with occasional summer occurrences in the Sandhills refuges but no regional breeding records, late April to early June and early August to late September.

Belted kingfisher. Common spring and fall migrant and widespread breeder (L, CL, FN, V), with post-1960 breeding records for Keya Paha, Brown, Lincoln, and Greeley Counties, mid-March to mid-November, overwintering frequently where open water is available.

Red-headed woodpecker. Common spring and fall migrant and occasional local breeder (FN) where large trees are present, with post-1960s breeding records for many Sandhills counties, early May to late September.

Red-bellied woodpecker. Occasional to rare permanent resident around the edges of the Sandhills (mainly the Niobrara and Platte Valleys), with post-1960 breeding records for Cherry, Lincoln, and Boone Counties.

Yellow-bellied sapsucker. Occasional to rare late fall and spring migrant or overwintering migrant, early October to mid-December and mid-March to late March.

Downy woodpecker. Common to uncommon permanent resident and local breeder (FN) in wooded areas, with post-1960 breeding records for many Sandhills counties.

Hairy woodpecker. Uncommon permanent resident and local breeder (L, FN) in wooded areas of river valleys or at the edges of the Sandhills, with post- 1960 breeding records for Keya Paha, Brown, Cherry, Keith, and Lincoln Counties.

Northern flicker. Common permanent resident and widespread breeder (L, FN, V) in wooded areas, with post-1960 breeding records for most Sandhills counties.

Olive-sided flycatcher. Uncommon spring and fall migrant, especially eastwardly, mid-May to late May and early to late September.

Western wood-pewee. Uncommon spring and fall migrant and local summer resident westwardly (occasional during summer at L, not proven to breed), with historic breeding records for the Niobrara Valley east to Cherry County, late May to early September. Eastern limits of the species and degrees of sympatric contact and possible hybridization with the following species are still uncertain.

Eastern wood-pewee. Uncommon spring and fall migrant and local breeder (FN), especially in eastern Niobrara Valley, mid-May to early September. There is a 1982 breeding record for Keya Paha County, which is probably near the western limit of this species' breeding range but where it nevertheless seems relatively common.

Yellow-bellied flycatcher. Uncommon to occasional spring and fall migrant eastwardly in deciduous thickets, rarely occurring west to Cherry County, early to late May and early to mid-September.

Alder flycatcher. Probably a rare spring and fall migrant, at least eastwardly, early to late May and mid-July to mid-September.

Willow flycatcher. Common to uncommon spring and fall migrant and local breeder (L, FN), mid-May to early September. Breeding limits are uncertain, owing to early confusion with the alder flycatcher, but the presence of singing males in the Niobrara Valley Preserve area suggest that breeding occurs there.

Least flycatcher. Common spring and fall migrant and probable local breeding resident (common breeder at Lacreek, not yet reported nesting in any Nebraska Sandhills refuges), early May to mid-May and late August to mid-September. No state breeding records since 1920, but apparently territorial (singing) males have been reported from Brown County.

Western flycatcher. Probably a very rare spring and fall migrant in the western Sandhills, with observations for Garden and McPherson Counties and a single state nesting record for Sioux County; too few records to indicate usual migration periods.

Eastern phoebe. Common spring and fall migrant and occasional local breeder (FN) in the Sandhills, with post-1960 breeding records for Keya Paha, Cherry, Brown, Lincoln, and Boone Counties, mid-April to late September.

Say's phoebe. Uncommon spring and fall migrant and local breeder (FN), especially westwardly, with post-1960 breeding records for Greeley, Lincoln, Keith, Garden, and Morrill Counties, mid-April to mid-September.

Great crested flycatcher. Uncommon spring and fall migrant and local breeder (FN, V) in wooded areas, especially in the Niobrara Valley, with post-1960s breeding records for Keya Paha, Brown, and Cherry Counties, late April to early September.

Cassin's kingbird. Rare spring and fall migrant and possible breeder in the northwestern Sandhills, where it sometimes occurs in dry ponderosa pine canyons, early May to mid-September. There are no documented Nebraska breeding records.

Eastern kingbird. Common to abundant spring and fall migrant and widespread breeder (L, CL, FN, V), with post-1960 breeding records for most Sandhills counties, early May to early September.

Western kingbird. Common to abundant spring and fall migrant and widespread breeder (L, CL, FN, V), with post-1960 breeding records for most Sandhills counties, early May to mid-September.

Horned lark. Common to abundant spring and fall migrant or permanent and widespread breeder (L, CL, FN, V), with breeding records for most Sandhills counties; spring and fall migrants from farther north supplement or replace the breeding population during winter.

Purple martin. Uncommon spring and fall migrant and very local breeder eastwardly, mostly in towns or villages, but largely absent from the Sandhills proper. There are post-1960 breeding records for Garden, Keith, and Lincoln Counties, mid-April to late August.

Tree swallow. Common spring and fall migrant and very local breeder (V) in wooded areas (nesting both in tree hollows and birdhouses), with post-1960 breeding records for Brown, Cherry, Sheridan, and Garden Counties, late April to mid-September.

Violet-green swallow. Uncommon to rare spring and fall migrant in the western Sandhills, with breeding limited to the Pine Ridge area of northwestern Nebraska (nesting in pine forests and birdhouses), mid-May to early June and early to late August.

Northern rough-winged swallow. Common spring and fall migrant and local breeder (FN, V) in rocky areas or other sites with natural cavities. There are post-1960 breeding records for Keya Paha, Brown, Cherry, Garden, and Keith Counties, late April to early September.

Bank swallow. Common spring and fall migrant and local breeder (L, V), with post-1960 breeding records for Sheridan, Garden, and Keith Counties, early May to early September. Seemingly the Sandhills should provide unlimited nesting habitats, but perhaps the sand is too unstable to provide good burrowing opportunities.

Barn swallow. Common spring and fall migrant and widespread breeder (L, CL, FN, V), with post-1960 breeding records for nearly all Sandhills counties, late April to late September.

Cliff swallow. Abundant spring and fall migrant and local breeder (L, FN), often under bridges or in culverts, with post-1960 breeding records for many Sandhills counties, late April to early September.

Blue jay. Common permanent resident and widespread breeder (CL, FN, V), with post-1960s breeding records for Brown, Cherry, Garden, Lincoln, and Greeley Counties.

Black-billed magpie. Uncommon permanent resident westwardly, becoming uncommon to occasional in the eastern Sandhills. Breeding locally (L, FN), often in cedar thickets, chokecherries, or other small trees; nesting records exist for many Sandhills counties.

American crow. Uncommon to occasional permanent resident and relatively widespread breeder (L, FN, V) where large trees are present. Post-1960 breeding records exist for Brown, Cherry, Sheridan, Keith, and Lincoln Counties.

Black-capped chickadee. Uncommon to occasional permanent resident and local breeder (L, FN) where large trees are present. Post-1960 breeding records exist for Keya Paha, Brown, Cherry, Keith, Lincoln, Dawson, and Boone Counties.

Red-breasted nuthatch. Uncommon permanent resident in northern Nebraska and local breeder in the Pine Ridge and Niobrara Valley at the edge of the Sandhills; an overwintering migrant elsewhere, early October to early April. Post-1960 breeding records exist for Keya Paha, Brown, and Cherry Counties.

White-breasted nuthatch. Uncommon permanent resident and local breeder (L, FN) in wooded areas, with post-1960 breeding records for Keya Paha and Cherry Counties.

Brown creeper. Uncommon or rare permanent resident, probably breeding locally in wooded areas of the Pine Ridge and Niobrara Valley, but with only a single, Brown County breeding record (nesting in a ponderosa pine in 1982); elsewhere in the Sandhills an overwintering migrant, mid-October to late March.

Rock wren. Uncommon to rare spring and fall migrant and local breeder in rocky locations at the edges of the Sandhills, with post-1960 breeding records for Keith and Lincoln Counties, early May to late October.

House wren. Common spring and fall migrant and widespread breeder (L, FN, V) in tree hollows or birdhouses, with post-1960 breeding records for nearly all Sandhills counties, late April to late September.

Winter wren. Occasional to rare spring and fall migrant and winter resident along shaded streams or canyons in Nebraska, mid-October to mid-April, but very rare in the Sandhills proper.

Sedge wren. Occasional to rare spring and fall migrant and possible local breeder (L, V) in the eastern and northern Sandhills, early to late May and early August to late September, but with no specific post-1960 breeding records for the region.

Marsh wren. Common to abundant spring and fall migrant and widespread breeder (L, CL, V) in wetlands, with post-1960 breeding records for Brown,

Cherry, Sheridan, Garden, Grant, and Keith Counties, early May to early October.

Golden-crowned kinglet. Uncommon to rare spring and fall migrant and winter resident, probably mostly in wooded areas at the northern edges of the Sandhills, mid-October to mid-April.

Ruby-crowned kinglet. Uncommon to occasional spring and fall migrant, mid-April to mid-May and late September to late October, probably mostly in wooded areas at the edges of the Sandhills; rare occurrences throughout the winter.

Eastern bluebird. Uncommon spring and fall migrant and local breeder (FN, V), with post-1960 breeding records for Rock, Brown, Lincoln, and Thomas Counties, late March to early November, but occasionally overwintering.

Mountain bluebird. Common spring and fall migrant and local breeder in the Pine Ridge, mid-March to mid-October, but probably not breeding in the Sandhills proper.

Townsend's solitaire. Common spring and fall migrant and winter resident, especially at western and northern edges of the Sandhills where junipers are common on rocky substrates, late September to late March.

Veery. Occasional to rare spring and fall migrant, mid-May to late May and late August to late September.

Gray-cheeked thrush. Uncommon to occasional spring and fall migrant, early to mid-May and mid-September to late October.

Swainson's thrush. Uncommon to occasional spring and fall migrant, early to late May and early to late September; has bred at least once in the Pine Ridge area and also in the Niobrara Valley (Brown County, 1900).

Hermit thrush. Uncommon to rare spring and fall migrant, mainly eastwardly, mid-April to late April and early to mid-October.

Wood thrush. Uncommon to occasional spring and fall migrant, mid-May to early June, and from late August to mid-September, and occasional or rare breeder. There are individual 1900–55 breeding records for Brown, Lincoln, and Logan Counties but no post-1960 records for the entire Sandhills region.

American robin. Common to abundant spring and fall migrant and widespread breeder (L, CL, FN, V), with post-1960 breeding records for nearly all Sandhills counties, late February to late November but frequently overwintering during mild winters.

Gray catbird. Uncommon spring and fall migrant and widespread breeder (L, FN, V), with post-1960 breeding records for Cherry, Lincoln, Dawson, and Greeley Counties, mid-May to late September.

Northern mockingbird. Occasional to rare spring and fall migrant and very local breeder (FN), with post-1960 breeding records for Cherry, Thomas, Keith, Lincoln, and Greeley Counties.

Brown thrasher. Common to uncommon spring and fall migrant and widespread breeder (L, CL, FN, V), with post-1960 breeding records for many Sandhills counties, late April to late September.

Water pipit. Common spring and fall migrant, especially near water, through-out the Sandhills, mid-April to late April and early October to late October.

Cedar waxwing. Common to uncommon spring and fall migrant, local or sporadic breeder, and an occasional overwintering migrant, late February to mid-March and early October to mid-December, but with frequent over-wintering in mild years. Occasional to uncommon at L and CL during summer, with no definite nesting records; post-1960 nesting records exist for Keith and Lincoln Counties and an early (pre-1920) record for Cherry County.

Northern shrike. Uncommon to occasional overwintering migrant, early November to mid-March.

Loggerhead shrike. Common to occasional spring and fall migrant and wide-spread breeder (L, CL, FN, V), with post-1960 breeding records for many Sandhills counties, early April to mid-September.

Bell's vireo. Uncommon to occasional spring and fall migrant and local breeder (L, CL, V), with post-1960 breeding records for Brown, Keith, and Lincoln Counties, mid-May to early September.

Solitary vireo. Occasional to rare spring and fall migrant and a rare breeder in the western Pine Ridge region (Sioux County), early to mid-May and mid-September to early October.

Yellow-throated vireo. Rare to very rare spring and fall migrant in the eastern Sandhills, accidental westwardly, early May to early June and early August to early September.

Warbling vireo. Common to uncommon spring and fall migrant and local breeder (L, CL, V), with post-1960 breeding records for Brown and Lincoln Counties, early May to early September.

Philadelphia vireo. Uncommon to rare spring and fall migrant (rarer west-wardly), mid-May to late May and late August to late September.

Red-eyed vireo. Uncommon to rare spring and fall migrant (rarer westwardly) and local breeder (L, FN, V), with post-1960 breeding records for Keya Paha and Brown Counties, mid-May to early September.

Tennessee warbler. Occasional to rare spring and fall migrant (rarer westwardly), early to late May and early September to early October. A bird with a brood patch was captured in Lincoln County in 1986, outside the species' normal breeding range.

Orange-crowned warbler. Common to uncommon spring and fall migrant, late April to mid-May and mid-September to mid-October.

Nashville warbler. Uncommon to occasional spring and fall migrant, early to mid-May and mid-September to early October.

Northern parula. Rare or accidental spring and fall migrant in the eastern Sandhills, with vagrant birds occurring west to Garden and Sheridan Counties, mid-April to mid-May and late August to late September.

Yellow warbler. Common spring and fall migrant and widespread breeder (L,

CL, FN, V), with post-1960 breeding records for Sheridan, Keith, Lincoln, Dawson, and Greeley Counties, early May to early September.

Magnolia warbler. Occasional to rare spring and fall migrant, mostly eastwardly, mid-May to late May and early September to early October.

Yellow-rumped warbler. Common spring and fall migrant, late April to mid-May and late September to late October, and local summer resident in the Pine Ridge area.

Black-throated green warbler. Rare to accidental spring and fall migrant in the Sandhills, mainly eastwardly but reported west to Lincoln and McPherson Counties, early to mid-May and mid-September to early October.

Blackburnian warbler. Rare spring and fall migrant, mainly eastwardly but reported west to Dawes County, early to mid-May and early September to early October.

Yellow-throated warbler. Rare to accidental spring and fall migrant, mainly eastwardly but observed west to Sheridan County, early to mid-May and early to late September.

Palm warbler. Rare to accidental spring and fall migrant, mainly eastwardly but reported west to Dawes County, early May and early October (though relatively few fall records exist).

Bay-breasted warbler. Rare to accidental spring and fall migrant, mainly eastwardly but reported west to the Panhandle, mid-May and mid-September to late September.

Blackpoll warbler. Uncommon to rare spring and fall migrant, mainly eastwardly, mid-May to late May and early to late September.

Black-and-white warbler. Uncommon to rare spring and fall migrant, mainly eastwardly, and a local breeder probably in the Niobrara Valley and the Pine Ridge valleys, early May to mid-September. Pre-1920 breeding records exist for Keya Paha, Brown, Cherry, and Thomas Counties, and there is also a 1957 record for Thomas County. Nesting very probably occurs in the Niobrara Valley Preserve, but proof is still lacking.

Cerulean warbler. Rare to accidental spring and fall migrant, mainly eastwardly but observed west to Keith and Sheridan Counties, early to mid-May and early to late August.

American redstart. Uncommon to occasional spring and fall migrant, mainly eastwardly, and local summer resident in the Pine Ridge and Niobrara Valley, mid-May to mid-September. There are pre-1960 breeding records for Keya Paha, Brown, Cherry, and Thomas Counties. Breeding may occur in the Bessey Division of Nebraska National Forest, Thomas County, and has been reported recently at V.

Ovenbird. Uncommon to rare spring and fall migrant and local breeder (FN, V) in the Niobrara Valley and Pine Ridge, with post-1960 breeding records for Keya Paha and Cherry Counties, early to mid-May and early to mid-September.

Northern waterthrush. Occasional to rare spring and fall migrant, early to mid-May and late August to late September.

Kentucky warbler. Very rare spring and fall migrant in eastern Nebraska, breeding in the southeastern part of the state mid-May to late August, but with no specific Sandhills records.

Connecticut warbler. Rare to accidental spring and fall migrant, mainly eastwardly, with records west to Cherry, Lincoln, and McPherson Counties, early to mid-May and mid-September to early October.

Mourning warbler. Rare spring and fall migrant in eastern Sandhills but reported west to Lincoln, Sheridan, and Scotts Bluff Counties.

Macgillivray's warbler. Rare spring and fall migrant in western Sandhills and probably a regular migrant in the Panhandle, early to mid-May and late August to mid-September.

Common yellowthroat. Common to abundant spring and fall migrant and widespread breeder (L, CL, FN, V), with post-1960 breeding records for most Sandhills counties, early May to mid-September.

Wilson's warbler. Occasional to rare spring and fall migrant, mainly eastwardly, mid-May to late May and early September to late September.

Canada warbler. Rare to very rare spring and fall migrant, mainly eastwardly but reported west to McPherson County, early to late May and early to mid-September.

Yellow-breasted chat. Uncommon spring and fall migrant and local breeder (L, FN), with post-1960 breeding records only for Dawes County (in 1981) but earlier records for Thomas and Logan Counties.

Scarlet tanager. Uncommon to rare spring and fall migrant, mainly eastwardly, and a very local breeder in the Niobrara Valley (one 1982 breeding record for Brown County), early May to late May and early August to mid-September.

Western tanager. Uncommon to rare spring and fall migrant, mainly westwardly, and a probable breeder in the Pine Ridge west of the Sandhills, mid-May to late May and early to late September.

Northern cardinal. Uncommon to occasional permanent resident and local breeder in southern and eastern edges of the Sandhills, with post-1960 breeding records for the counties of the North Platte and Platte Valleys and for Boone County. There are no recent nesting records for the Niobrara Valley, but territorial pairs are common as far west as Brown County.

Rose-breasted grosbeak. Uncommon spring and fall migrant and local breeder eastwardly at the edges of the Sandhills, early May to mid-September. There are early breeding records west to Brown County but no post-1960 breeding records for the Sandhills counties.

Black-headed grosbeak. Uncommon spring and fall migrant and local breeder (FN), especially westwardly, with post-1960 breeding records for Keya Paha, Brown, Cherry, and Lincoln Counties, mid-May to late August; sometimes hybridizes with preceding species.

Blue grosbeak. Uncommon spring and fall migrant and local breeder (CL, FN, V), with post-1960 breeding records for McPherson, Lincoln, and Greeley Counties, late May to late August.

Lazuli bunting. Uncommon spring and fall migrant and local breeder westwardly (v), mid-May to late August; sometimes hybridizes with the following species.

Indigo bunting. Uncommon spring and fall migrant and local breeder eastwardly, with post-1960 breeding records for Keya Paha and Brown Counties, mid-May to late August.

Dickcissel. Common to occasional spring and fall migrant and widespread but seemingly erratic breeder (L, CL, FN, V), with post-1960 breeding records for Garden, McPherson, Lincoln, and Greeley counties (more common eastwardly), mid-May to late August.

Rufous-sided towhee. Uncommon to occasional spring and fall migrant and local breeder (FN, V), with post-1960 breeding records for Garden, McPherson, and Lincoln Counties, April to mid-October.

American tree sparrow. Common overwintering migrant, late October to early April.

Chipping sparrow. Common spring and fall migrant and locally common breeder (FN, V), with post-1960 breeding records for Keya Paha, Brown, Cherry, and Boone Counties, late April to early October.

Clay-colored sparrow. Common spring and fall migrant and very local breeder northwestwardly, early to mid-May and early September to early October. There are pre-1960 breeding records for Cherry and Sheridan Counties but no known state nesting since 1973.

Field sparrow. Common spring and fall migrant and locally common breeder (FN, V), with post-1960 breeding records for Keya Paha, Brown, Cherry, Keith, McPherson, and Lincoln Counties, late April to early October.

Vesper sparrow. Uncommon spring and fall migrant and widespread breeder (L, CL, FN, V), with post-1960 breeding records for McPherson and Lincoln Counties, mid-April to early October.

Lark sparrow. Common to abundant spring and fall migrant and widespread breeder (L, CL, FN, V), with post-1960 breeding records for most Sandhills counties, early May to early September.

Lark bunting. Common spring and fall migrant and widespread breeder, especially in the western Sandhills (L, CL, NF, V), with post-1960 breeding records for Cherry, Box Butte, Garden, Arthur, Keith, McPherson, Lincoln, and Greeley Counties, mid-May to late August.

Savannah sparrow. Uncommon to occasional spring and fall migrant and possible local breeder in northwestern Nebraska, with nesting reported for FN and summering without proven nesting at V, CL, and L; no specific recent nesting records for the state.

Baird's sparrow. Uncommon to occasional spring and fall migrant, early April to early May and late September to mid-October.

Grasshopper sparrow. Common to abundant spring and fall migrant and widespread breeder (CL, FN, V), with post-1960 breeding records for Brown, Garden, Arthur, Keith, McPherson, and Thomas Counties, early May to early September.

LeConte's sparrow. A probably rare but inconspicuous spring and fall migrant through the Sandhills, late April to early May and late September to late October.

Sharp-tailed sparrow. An apparently rare but very inconspicuous spring and fall migrant, mainly May and October, but with few specific records.

Fox sparrow. An uncommon spring and fall migrant, late March to mid-April and mid-October to mid-November.

Song sparrow. A common spring and fall migrant but seemingly rather local breeder (L, V), with post-1960 breeding records for McPherson and Boone Counties.

Lincoln's sparrow. A common spring and fall migrant, late April to mid-May and mid-September to mid-October.

Swamp sparrow. An uncommon spring and fall migrant and very local breeder (L, CL), with a single post-1960 breeding record for Rock County but probably nesting locally in the marshes of Keith, Garden, and Sheridan Counties, late April to early May and late September to late October.

White-throated sparrow. A common overwintering migrant, early October to mid-May.

White-crowned sparrow. A common overwintering migrant, early October to mid-May.

Harris' sparrow. A common overwintering migrant, mid-October to mid-May.

Dark-eyed junco. A common overwintering migrant, early October to mid-April.

McCown's longspur. A common overwintering migrant eastwardly, October to April; breeding in the Panhandle west of the Sandhills, early April to early October.

Lapland longspur. A common overwintering migrant, mid-November to late February.

Chestnut-collared longspur. A common overwintering migrant eastwardly, October to April; breeding in the Panhandle just west of the Sandhills and in southern South Dakota, mid-April to early October.

Snow bunting. A common overwintering migrant, mid-November to mid-February.

Bobolink. A common spring and fall migrant and widespread breeder (L, CL, FN, V), with post-1960 breeding records for Rock, Sheridan, Garden, Grant, Keith, and Lincoln Counties, mid-May to mid-August.

Red-winged blackbird. An abundant spring and fall migrant and widespread breeder (L, CL, FN, V), with post-1960 breeding records for nearly all Sandhills counties, early March to late November.

Eastern meadowlark. A common spring and fall migrant and breeder in the eastern Sandhills, but with local breeding west to the Panhandle (L, CL,

FN, V), especially in mesic areas such as low meadows, early April to mid-October, but occasionally overwintering in mild years. Post-1960 breeding records exist only for Sheridan County in the Sandhills region.

Western meadowlark. An abundant spring and fall migrant and widespread breeder (L, CL, FN, V), with post-1960 breeding records for most Sandhills counties, early March to late October, but commonly overwintering in mild years.

Yellow-headed blackbird. A common, abundant spring and fall migrant and locally common breeder (L, CL, FN, V), with post-1960 breeding records for Cherry, Sheridan, Garden, Grant, McPherson, and Lincoln Counties, late April to mid-September.

Rusty blackbird. An uncommon spring and fall migrant, late March to mid-April and early November to late December, but probably overwintering occasionally.

Brewer's blackbird. A common spring and fall migrant, especially westwardly, with breeding occurring in the Panhandle west of the Sandhills and at L and reportedly also V, mid-April to mid-November. Post-1960 breeding records exist only for Morrill County in the Sandhills region.

Common grackle. An abundant spring and fall migrant and widespread breeder (L, CL, FN, V), with post-1960 breeding records for most Sandhills counties, late March to late October.

Brown-headed cowbird. An abundant spring and fall migrant and widespread breeder (L, CL, FN, V), with post-1960 breeding records for many Sandhills counties, mid-April to early October.

Orchard oriole. A common spring and fall migrant and widespread breeder (L, CL, FN, V), with post-1960 breeding records for most Sandhills counties, early May to late August.

Northern oriole. A common spring and fall migrant and widespread breeder (CL, FN, V), with post-1960 breeding records for Rock, Brown, Keith, McPherson, Lincoln, Dawson, and Greeley Counties, early May to early September; mainly present as the "Baltimore" (eastern) race but with "Bullock's" orioles or hybrid phenotypes increasingly present westwardly.

Purple finch. An uncommon overwintering migrant, especially toward the Panhandle, late October to late April.

House finch. An increasingly common permanent resident in the Platte Valley and villages or cities along the southern edge of the Sandhills, spreading east across the state along the Platte Valley during the 1980s.

Red crossbill. An irregular overwintering migrant, especially toward the Panhandle, mid-November to early April.

Common redpoll. An occasional overwintering migrant, late November to mid-March.

Pine siskin. A common overwintering migrant and a breeder in the Pine Ridge, mid-October to mid-May.

American goldfinch. A common permanent resident and widespread breeder (L, CL, FN, V), with post-1960 breeding records for many Sandhills counties.

Evening grosbeak. A rare overwintering migrant, especially toward the Panhandle, early November to late April.

House sparrow. A common to abundant introduced permanent resident.

Sources: Based mainly on Johnsgard 1986, Ducey 1988, and checklists for national wildlife refuges (summarized in Jones 1990). Sandhills county breeding records are mostly after Ducey 1988, although nearly all these counties also include some non-Sandhills habitats (see table 4). Times of occurrence normally based on median arrival and departure dates as reported in Johnsgard 1986. Additional regional information for northwestern Nebraska (Pine Ridge area) based on Rosche 1982. Accidental and extirpated species excluded. Order of entries is based on the American Ornithologists Union's 1983 *Checklist.*

Central Sandhills Vascular Plant Taxa

FAMILIES, GENERA, AND NUMBER OF SPECIES

Aceraceae (maple family): *Acer* 1

Aizoaceae (carpetweed family): *Mollugo* 1

Alismataceae (water plantain family): *Alisma* 1, *Sagittaria* 2

Amaranthaceae (pigweed family): *Amaranthus* 7, *Froelichia* 1

Anacardiaceae (cashew family): *Rhus* 2

Apiaceae (parsley family): *Berula* 1, *Cicuta* 2, *Osmorhiza* 1, *Sanicula* 2, *Sium* 1

Apocynaceae (dogbane family): *Apocynum* 1

Asclepiaceae (milkweed family): *Asclepias* 10

Asteraceae (composite family): *Achillea* 1, *Ambrosia* 2, *Antennaria* 1, *Artemisia* 4, *Aster* 5, *Bidens* 5, *Chrysopsis* 2, *Cirsium* 3, *Coreopsis* 1, *Crepis* 1, *Dyssodia* 1, *Erigeron* 2, *Eupatorium* 2, *Grindelia* 1, *Gutierrezia* 1, *Happlopappus* 1, *Helenium* 1, *Helianthus* 5, *Hymenopappus* 2, *Iva* 1, *Kuhnia* 1, *Lactuca* 3, *Liatris* 2, *Lygodesmia* 2, *Machaeranthera* 1, *Microseris* 1, *Ratibida* 1, *Rudbeckia* 1, *Senecio* 4, *Solidago* 9, *Thelesperma* 1, *Townsendia* 1, *Tragopogon* 1, *Xanthium* 1

Balsaminaceae (touch-me-not family): *Impatiens* 1

Boraginaceae (borage family): *Cryptantha* 3, *Hackelia* 1, *Lappula* 2, *Lithospermum* 2, *Onosmodium* 1

Brassicaceae (mustard family): *Arabis* 1, *Berteroa* 1, *Brassica* 1, *Camelina* 1, *Capsella* 1, *Cardamine* 1, *Descurainia* 1, *Erysimum* 2, *Lepidium* 2, *Lesquerella* 1, *Nasturtium* 1, *Rorippa* 1, *Sisymbrium* 2, *Thlaspi* 1

Cactaceae (cactus family): *Coryphantha* 2, *Opuntia* 3

Campanulaceae (bellflower family): *Campanula* 1, *Lobelia* 2, *Triodanis* 1

Cannabaceae (hemp family): *Humulus* 1

Capparaceae (caper family): *Cleome* 1, *Cristatella* 1, *Polanisia* 1

Caprifoliaceae (honeysuckle family): *Symphoricarpus* 1

Caryophyllaceae (pink family): *Arenaria* 1, *Lychnis* 2, *Silene* 1, *Stellaria* 1, *Vaccaria* 1

Celastraceae (staff-tree family): *Celastrus* 1

Ceratophyllaceae (hornwort family): *Ceratophyllum* 1

Chenopodiaceae (goosefoot family): *Atriplex* 1, *Ceratoides* 1, *Chenopodium* 5, *Corispermum* 2, *Cycloloma* 1, *Kochia* 1, *Salsola* 1

Commelinaceae (spiderwort family): *Commelina* 1, *Tradescantia* 2

Convolvulaceae (convolvulus family): *Convolvulus* 1, *Ipomoea* 1

Cornaceae (dogwood family): *Cornus* 1

Cupressaceae (cypress family): *Juniperus* 2

Cuscutaceae (dodder family): *Cuscuta* 1

Cyperaceae (sedge family): *Carex* 24, *Cyperus* 5, *Eleocharis* 4, *Scirpus* 6

Elatinaceae (waterwort family): *Elatine* 1

Equisetaceae (horsetail family): *Equisetum* 3

Euphorbiaceae (spurge family): *Croton* 1, *Euphorbia* 7

Fabaceae (legume family): *Amorpha* 1, *Amphicarpa* 1, *Apios* 1, *Astragalus* 6, *Dalea* 1, *Desmodium* 1, *Glycyrrhiza* 1, *Lathyrus* 1, *Lespedeza* 1, *Lotus* 1, *Medicago* 1, *Melilotis* 1, *Oxytropis* 1, *Petalostemom* 3, *Psoralea* 4, *Robinia* 1, *Strophostyles* 1, *Trifolium* 1, *Vicia* 2

Fagaceae (beech family): *Quercus* 1

Gentianaceae (gentian family): *Gentiana* 1

Haloragaceae (water milfoil family): *Myriophyllum* 1

Hippuridacea (mare's-tail family): *Hippuris* 1

Hydrocharitaceae (frog's-bit family): *Elodea* 1

Hydrophyllaceae (waterleaf family): *Ellisia* 1

Hypericaceae (St. John's–wort family): *Hypericum* 1

Iridaceae (iris family): *Sisyrinchium* 1

Juncaceae (rush family): *Juncus* 7

Juncaginaceae (arrowgrass family): *Triglochin* 2

Lamiaceae (mint family): *Hedeoma* 2, *Lycopus* 3, *Mentha* 1, *Monarda* 2, *Prunella* 1, *Salvia* 1, *Scutellaria* 2, *Stachys* 1

Lemnaceae (duckweed family): *Lemna* 4, *Spirodela* 1, *Wolffia* 2

Lentibulariaceae (bladderwort family): *Utricularia* 1

Liliaceae (lily family): *Allium* 2, *Leucocrinum* 1, *Lilium* 1, *Polygonatum* 1, *Smilacina* 1, *Smilax* 1, *Yucca* 1

Linaceae (flax family): *Linum* 2

Loasaceae (sand lily family): *Mentzelia* 2

Lythraceae (loosestrife family): *Lythrum* 1

Malvaceae (mallow family): *Callirhoe* 1, *Sphaeralcea* 1

Marsileaceae (marsilea family): *Marsilea* 1

Najadaceae (naiad family): *Najas* 2

Nyctaginaceae (four-o'clock family): *Abronia* 1, *Mirabilis* 3

Nymphaeaceae (water lily family): *Nuphar* 1

Oleaceae (olive family): *Fraxinus* 1

Onagraceae (evening primrose family): *Calylophus* 1, *Circaea* 1, *Epilobium* 2, *Gaura* 2, *Oenothera* 6.

Ophioglossaceae (adder's-tongue family): *Botrychium* 1

Orchidaceae (orchid family): *Habenaria* 2, *Spiranthes* 1

Orobanchaceae (broomrape family): *Orobanche* 1

Oxalidaceae (wood sorrel family): *Oxalis* 1

Plantaginaceae (plantain family): *Plantago* 1

Poaceae (grass family): *Agrohordeum* 1, *Agropyron* 5, *Agrostis* 13, *Alopecurus* 1, *Andropogon* 3, *Aristida* 3, *Beckmannia* 1, *Bouteloua* 3, *Bromus* 8, *Buchloe* 1, *Calamogrostis* 2, *Calamovilfa* 1, *Cenchrus* 1, *Digitaria* 1, *Distichlis* 1, *Echinochloa* 1, *Elymus* 3, *Eragrostis* 4, *Festuca* 1, *Glyceria* 2, *Hordeum* 2, *Koeleria* 1,

Leptochloa 1, *Muhlenbergia* 3, *Munroa* 1, *Oryzopsis* 1, *Panicum* 5, *Paspalum* 1, *Phalaris* 1, *Phleum* 1, *Phragmites* 1, *Poa* 4, *Polypogon* 1, *Redfieldia* 1, *Schedonnardus* 1, *Schizachne* 1, *Scholochloa* 1, *Setaria* 1, *Sorghastrum* 1, *Spartina* 2, *Sphenopholis* 1, *Sporobolus* 1, *Stipa* 3

Polemoniaceae (polemonium family): *Ipomopsis* 1, *Phlox* 1

Polygonaceae (buckwheat family): *Eriogonum* 3, *Fagopyrum* 1, *Polygonum* 10, *Rumex* 5

Polypodiaceae (fern family): *Cystopteris* 1, *Dryopteris* 2, *Onoclea* 1, *Thelypteris* 1, *Woodsia* 1

Portulacaceae (purslane family): *Portulaca* 1, *Talinum* 1

Potamogetonaceae (pondweed family): *Potamogeton* 10

Primulaceae (primrose family): *Lysimachia* 2

Ranunculaceae (buttercup family): *Anemone* 3, *Clematis* 1, *Delphinium* 1, *Ranunculus* 8, *Thalicrum* 1

Rhamnaceae (buckthorn family): *Ceanothus* 1, *Rhamnus* 1

Rosaceae (rose family): *Agrimonia* 2, *Crataegus* 1, *Fragaria* 1, *Geum* 1, *Potentilla* 6, *Prunus* 3, *Rosa* 2, *Rubus* 1

Rubiaceae (madder family): *Galium* 4

Saliceae (willow family): *Populus* 1, *Salix* 6

Salviniaceae (salvinia family): *Azolla* 1

Santalceae (sandalwood family): *Comandra* 1

Saxifragiceae (saxifrage family): *Ribes* 2

Scrophulariaceae (figwort family): *Agalinis* 2, *Castilleja* 1, *Mimulus* 2, *Penstemon* 5, *Veronica* 2

Selaginellaceae (selaginella family): *Selaginella* 1

Solanaceae (nightshade family): *Datura* 1, *Physalis* 4, *Solanum* 3

Sparganiaceae (bur-reed family): *Sparganium* 2

Typhaceae (cattail family): *Typha* 2

Ulmaceae (elm family): *Celtis* 1, *Ulmus* 1

Urticaceae (nettle family): *Parietaria* 1, *Urtica* 1

Verbenaceae (vervain family): *Verbena* 4

Violaceae (violet family): *Viola* 3

Vitaceae (grape family): *Parthenocissus* 1, *Vitis* 2

Zanichelliaceae (horned pondweed family): *Zannichellia* 1

Zygophyllaceae (caltrop family): *Tribulus* 1

Source: Plant taxa are those reported in Great Plains Flora Assoc. 1977 for the region encompassed by Grant, Hooker, Thomas, Arthur, McPherson, and Logan Counties, excluding species listed as rare or restricted.

TAXONOMIC AND ECOLOGICAL SUMMARY OF
SANDHILLS FLORA

	Six Counties	Keith County
Total area (sq. mi.)	4,349	1,039
Total area (km²)	11,263	2,810
Area within Sandhills	c. 95%	c. 40%
Total plant families	87	87
Total genera	271	302
Total species (spp.)	454+[a]	c. 599[a]
Spp. totals, major families		
Poaceae	91 (20.0%)	98 (16.4%)
Asteraceae	71 (15.6%)	97 (16.2%)
Cyperaceae	39 (8.6%)	33 (5.5%)
Fabaceae	30 (6.6%)	44 (7.3%)
Polygonaceae	19 (4.2%)	18 (3.0%)
Brassiceae	17 (3.7%)	20 (3.3%)
Rosaceae	17 (3.7%)	11 (1.8%)
Ranunculaceae	14 (3.1%)	7 (1.2%)
Lamiaceae	13 (2.9%)	15 (2.5%)
Onagraceae	12 (2.6%)	17 (2.8%)
Scrophulariaceae	12 (2.6%)	15 (2.5%)
Chenopodiaceae	12 (2.6%)	13 (2.2%)
Growth-habit distribution		
Herbaceous perennials	—	58.1%
Herbaceous annuals	—	30.4%
Woody perennials	—	7.5%
Biennials	—	4.0%
Relative habitat distribution[b]		
Disturbance spp. (ruderals)	—	43.6%
Mixed/shortgrass prairie spp.	—	26.5%
Riparian habitat spp.	—	26.5%
Sandhills prairie spp.	—	22.2%
Lowland/prairie meadow spp.	—	22.0%
Woodland habitat spp.	—	9.0%
Aquatic habitat spp.	—	8.7%
Other habitats	—	5.7%

Source: Keith County data from Sutherland and Rolfsmeier (1989).

[a]The totals for Keith County include all rare species, as well as subspecies and varieties; those for the six counties do not. If one adjusts for these differences by eliminating all multiple subspecies and varieties, the total species-level taxa for Keith County are reduced to 588; however, about 20 additional species have been found there since 1989 (S. B. Rolfsmeier, pers. comm.).

[b]Habitat distributions total more than 160% because many species occur in multiple habitats. About 1.9% of Keith County is in woodlands; for the Sandhills as a whole this percentage is about 1.8% (Schmidt 1986).

Parks, Forests, Nature Preserves, and Other Natural Areas of the Sandhills and Vicinity

Arapaho Prairie. This Sandhills prairie area, located in Arthur County, is owned by the Nature Conservancy and managed for research and conservation purposes by the University of Nebraska at Lincoln. It consists of 1,298 acres (526 hectares) of upland prairie. Various floral (Keeler, Harrison, and Vescio 1980) and faunal (Ballinger, Lynch, and Cole 1979; Joern 1982) checklists have been published. *See also* Cedar Point Biological Station.

Arthur Bowring Sandhills Ranch Historical State Park. Located in northern Cherry County, near Merriman, this recently established (1988) park and still-functioning homesteaded ranch of 7,200 acres (2,900 hectares) includes much native Sandhills grassland. For information, write Nebraska Game and Parks Commission, P.O. Box 30370, Lincoln NE 68503.

Ashfall Fossil Beds State Historical Park. This relatively undeveloped area of 360 acres (146 hectares) is located in the Verdigre Creek Valley of Antelope County, six miles (nearly ten kilometers) north of Royal, Nebraska. The remarkably preserved 10-million-year-old exposed fossil site, Poison Ivy Quarry, is of particular interest to paleontologists, but the park includes some native prairie habitat as well. For information, write Park Manager, P.O. Box 66, Royal NE 68773.

Bessey Division, Nebraska National Forest. This planted forest of various pines and the adjoining areas of Sandhills grasslands and riverine forests lies in the central Sandhills in eastern Thomas County. Of its 90,450 acres (36,620 hectares), 22,000 acres (8,900 hectares) are planted to forest. A bird checklist of 95 species, including 36 breeders, and related information are available from Superintendent, Halsey NE 69142.

Calamus State Recreation Area. This area of 11,200 acres (4,530 hectares) near Burwell includes about 6,000 acres (2,400 hectares) of Sandhills habitat surrounding Calamus Reservoir in Garfield and Loup Counties.

Cedar Point Biological Station. This facility, owned by the University of Nebraska and operated for teaching and research by the university's School of Biological Sciences, is located in Keith County, just south of the Sandhills boundary. It lies along the southern shoreline of Lake Ogallala, a 640-acre (260-hectare) impoundment directly east of Lake McConaughy. The station consists of about 37.5 acres (15 hectares) and is situated at the base of a canyon largely dominated by mixed grasses and juniper woodlands; close by are many stands of riparian deciduous woodlands. Arapaho Prairie (see above), a mostly virgin Sandhills prairie 17 miles to the north, is a major focus of ecological research. Much biological information has been gathered at the station and about 200 publications or unpublished dissertations and theses have been produced by its faculty and students since it was established in 1975. For a bird list of 244 species recorded in the station's vicinity and along the North Platte Valley west to Lewellen, see Rosche and Johnsgard 1984. For a botanical list of the approximately 600 taxa (species, subspecies, and varieties) of vascular plants known from Keith County, see Sutherland and Rolfsmeier 1989. Information on the biological station and its diverse activities can be obtained from Station Director, School of Biological Sciences, University of Nebraska, Lincoln NE 68588-0118.

Crescent Lake National Wildlife Refuge. This refuge, 28 miles (45 kilometers) north of Oshkosh and 28 miles south of Lakeside, in Garden County, covers 46,000 acres (18,600 hectares). It includes Sandhills grasslands (95 percent of the total refuge area), about 13 variably alkaline and relatively shallow lakes, and several additional small wetlands. A checklist of 218 bird species, including 64 breeders, and related information are available from Refuge Manager, Star Route Box 31, Ellsworth NE 69341.

Fort Niobrara National Wildlife Refuge. Situated three miles (about five kilometers) east of Valentine in Cherry County, this refuge contains 19,100 acres (7,700 hectares) and consists mostly (60 percent) of Sandhills prairie, with some mixed hardwoods along the Niobrara River. Most of the land is primarily used as bison range. Part of the refuge—about 4,600 acres (1,850 hectares) to the north of the Niobrara River—has been designated as the Fort Niobrara Wilderness Area. A bird checklist of 201 species, including 76 breeders, and related information are available from Refuge Manager, Hidden Timber Route, Valentine NE 69201.

Graves Ranch Preserve. This small 840-acre (340-hectare) area near Crescent Lake National Wildlife Refuge was established in 1984 specifically to protect the endangered and blowout-adapted Hayden's penstemon.

Grove Lake Wildlife Management Area. This small and undeveloped area of 1,500 acres (600 hectares) of Sandhills prairie and bur oak woodland is located two miles (three kilometers) north of Royal, in Antelope County. For information, write Nebraska Game and Parks Commission, P.O. Box 30370, Lincoln NE 68503.

Gudmundsen Sandhills Laboratory. This 12,817-acre (5,190-hectare) area of Sandhills grassland is located north of Whitman in Grant, Cherry, and Hooker Counties. It is owned by the University of Nebraska Foundation and operated by the university's Institute for Agriculture and Natural Resources (IANR) for ranch-related research. Information can be obtained from the IANR, University of Nebraska, Lincoln NE 68588.

Lacreek National Wildlife Refuge. Located in Bennett County, South Dakota, just north of the Nebraska border and about ten miles northeast of Merriman (Cherry County), this refuge consists of about 9,400 acres (3,800 hectares), including upland Sandhills prairie (40 percent of total area), lowland meadows, and some 5,000 acres (2,000 hectares) of shallow wetlands (mostly impounded pools). A bird checklist of 235 species observed at the refuge, including 91 breeding species, is available from Refuge Manager, Martin SD 57551.

Long Lake State Recreation Area. This undeveloped area in southern Brown County includes 80 acres (32 hectares) of typical Sandhills wetland and prairie habitats.

Niobrara River Sanctuary. This small prairie and woodland sanctuary, owned by the National Audubon Society, consists of 218 acres (88 hectares) lying along a half-mile stretch of riverfront on the south bank of the Niobrara, in Rock County. Access is by permission only.

Niobrara Valley Preserve. This national area of mixed grassland and forested valley habitats lies in the ecologically and evolutionarily important transition zone between western coniferous and eastern hardwood forests. It includes ponderosa pine, eastern deciduous forest, northern boreal forest, Sandhills prairie, mixed-grass prairie, and tallgrass prairie habitats. It extends east from the Fort Niobrara National Wildlife Refuge in Cherry County into western Keya Paha County, and comprises 54,000 acres (22,000 hectares) of land owned and managed by the Nature Conservancy. In 1991 about 70 miles (110 kilometers) of this stretch of river was added to the list of nationally recognized Wild and Scenic Rivers. There is not yet a complete bird checklist available, but Brogie and Mossman (1983) published a breeding list of at least 105 species for the preserve, and Kaul, Kantak, and Churchill (1988) have summarized biogeographic information for the general Niobrara Valley region. Information on the preserve is available from the Conservancy's Nebraska Office, 1722 St. Mary's, Omaha NE 68102, or Preserve Manager, Rte. 1, Johnstown NE 69214.

Oglala National Grassland. This area of 94,000 acres (38,000 hectares) in northern Sioux and northwestern Dawes Counties is one of 19 national grasslands administered by the U.S. Forest Service. It consists of shortgrass prairie and badlands topography, including Toadstool Geologic Park.

Pine Ridge region. This large area of the northern Panhandle in Sioux and Dawes Counties includes 49,900 acres (20,200 hectares) in the Pine Ridge District of the Nebraska National Forest, plus the Oglala National Grassland (see above), both administered by the U.S. Forest Service; Fort Robinson State Park (22,000 acres; 8,900 hectares) and Chadron State Park (972 acres; 390 hectares), both administered by the Nebraska Game and Parks Commission; and the undeveloped 6,000-acre (2,400-hectare) Pine Ridge National Recreational Area. The Forest Service has a regional bird checklist of 302 species. Information on the federal lands of the Pine Ridge region is available from Forest Supervisor, U.S. Forest Service, 270 Pine St., Chadron NE 69337. The Nebraska Game and Parks Commission, P.O. Box 30370, Lincoln NE

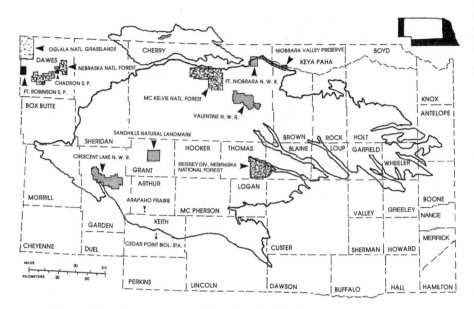

Fig. 48. Location of counties, plus selected wildlife refuges, preserves, parks, and other protected areas in the Sandhills region.

68503, can provide information on the region's two state parks, as can their superintendents: Ft. Robinson State Park, P.O. Box 392, Crawford NE 69339, and Chadron State Park, Chadron NE 69337.

Pressey Wildlife Management Area. This small and undeveloped area of 1,500 acres (600 hectares) of mixed prairie, riparian woodland, and wetlands is located near the South Loup River, five miles (eight kilometers) northeast of Oconta in Custer County. For information, write Nebraska Game and Parks Commission, P.O. Box 30370, Lincoln NE 68503.

Samuel R. McKelvie National Forest. Previously Niobrara Division, Nebraska National Forest, this preserve consists of 115,700 acres (46,850 hectares), including 2,300 acres (930 hectares) planted to trees. It is located in central Cherry County and was the site of a forest nursery until 1925. Historical information on the Nebraska National Forest is summarized in Hunt 1965.

Sandhills National Natural Landmark. This still-undeveloped area consists of 32,900 acres (13,300 hectares) of Sandhills grassland in central Grant County. No faunal checklists or other sources of biological information are yet available.

Smith Falls State Park. This recently established park, located 18 miles east of Valentine in Cherry County, is within the Niobrara Wild and Scenic River area. Its 200 acres (81 hectares) contain the state's largest waterfall, whose waters empty into the south side of the Niobrara, as well as typical deciduous springbranch canyon vegetation.

Valentine National Wildlife Refuge. This refuge is about 20 miles south of Valentine in Cherry County. Its more than 71,000 acres (29,000 hectares) consists mostly (85 percent) of Sandhills prairie but includes about 10,000 acres (4,000 hectares) of associated wetlands (some 25 shallow lakes and additional smaller marshes). A bird checklist of 221 species, including 93 breeders, and related information are available from Refuge Manager, Hidden Timber Route, HC Box 67, Valentine NE 69201.

Vernacular and Scientific Names of Plants and Animals Mentioned in the Text (Sandhills and Elsewhere)

MAMMALS

Badger, *Taxidea taxus*
Beaver, *Castor canadensis*
Big brown bat, *Eptesicus fuscus*
Bison, *Bison bison*
Black-footed ferret, *Mustela nigripes*
Black-tailed jackrabbit, *Lepus californicus*
Black-tailed prairie dog, *Cynomys ludovicianus*
Bobcat, *Lynx rufus*
Buffalo, *see* bison
Cactus mouse, *Peromyscus eremicus*
Coyote, *Canis latrans*
Deer mouse, *Peromyscus maniculatus*
Desert cottontail, *Sylvilagus audubonii*
Eastern cottontail, *Sylvilagus floridanus*
Eastern mole, *Scalopax aquaticus*
Eastern spotted skunk, *Spilogale putorius*
Eastern woodrat, *Neotoma floridana*
Elk, *Cervus canadensis*
Fox squirrel, *Sciurus niger*
Franklin's ground squirrel, *Spermophilus franklinii*
Fringe-tailed bat, *Myotis thysanodes*
Fulvous harvest mouse, *Reithrodontomys fulvescens*
Gray wolf, *Canis lupus*
Hispid cotton rat, *Sigmodon hispidus*
Hispid pocket mouse, *Perognathus hispidus*
Hoary bat, *Lasiurus cinereus*
Keen's bat (or myotis), *Myotis keenii*
Eastern chipmunk, *Tamias striatus*
Least shrew, *Cryptotis parva*
Least weasel, *Mustela nivalis*
Long-tailed weasel, *Mustela frenata*
Masked shrew, *Sorex cinereus*
Meadow jumping mouse, *Zapus hudsonius*

Meadow vole, *Microtus pennsylvanicus*
Merriam's kangaroo rat, *Dipodomys merriami*
Mink, *Mustela vison*
Mule deer, *Odocoileus hemionus*
Muskrat, *Ondatra zibethicus*
Northern grasshopper mouse, *Onychomys leucogaster*
Northern pocket gopher, *Thomomys talpoides*
Olive-backed pocket mouse, *Perognathus fasciatus*
Opossum, *Didelphis virginiana*
Ord's kangaroo rat, *Dipodomys ordii*
Plains pocket gopher, *Geomys bursarius*
Plains harvest mouse, *Reithrodontomys montanus*
Plains pocket mouse, *Perognathus flavescens*
Porcupine, *Erethizon dorsatum*
Prairie vole, *Microtus ochrogaster*
Pronghorn, *Antilocapra americana*
Raccoon, *Procyon lotor*
Red bat, *Lasiurus borealis*
Red fox, *Vulpes vulpes*
Richardson's ground squirrel, *Spermophilus richardsoni*
Short-tailed shrew, *Blarina brevicauda*
Silky pocket mouse, *Perognathus flavus*
Silver-haired bat, *Lasionycteris noctivagans*
Small-footed bat (or myotis), *Myotis leibii*
Southern bog lemming, *Synaptomys cooperi*
Southern grasshopper mouse, *Onychomys torridus*
Spotted ground squirrel, *Spermophilus spilosoma*
Striped skunk, *Mephitis mephitis*
Swift fox, *Vulpes velox*
Thirteen-lined ground squirrel, *Spermophilus tridecemlineatus*
Western harvest mouse, *Reithrodontomys megalotis*
White-footed mouse, *Peromyscus leucopus*
White-tailed deer, *Odocoileus virginianus*
White-tailed jackrabbit, *Lepus townsendii*

BIRDS

Acadian flycatcher, *Empidonax virescens*
Alder flycatcher, *Empidonax alnorum*
American avocet, *Recurvirostra americana*
American bittern, *Botaurus lentiginosus*
American black duck, *Anas rubripes*
American coot, *Fulica americana*
American crow, *Corvus brachyrhynchos*

American goldfinch, *Carduelis tristis*
American kestrel, *Falco sparverius*
American redstart, *Setophaga ruticilla*
American robin, *Turdus migratorius*
American tree sparrow, *Spizella arborea*
American white pelican, *Pelecanus erythrorhynchos*
American wigeon, *Anas americana*
American woodcock, *Scolopax minor*
Baird's sandpiper, *Calidris melanotos*
Baird's sparrow, *Ammodramus bairdii*
Bald eagle, *Haliaeetus leucocephalus*
Baltimore oriole, *see* northern oriole
Bank swallow, *Riparia riparia*
Barn swallow, *Hirundo rustica*
Barred owl, *Strix varia*
Bay-breasted warbler, *Dendroica castanea*
Bell's vireo, *Vireo bellii*
Belted kingfisher, *Ceryle alcyon*
Black-and-white warbler, *Dendroica varia*
Black-bellied plover, *Pluvialis squatarola*
Black-billed cuckoo, *Coccyzus erythropthalmus*
Black-billed magpie, *Pica pica*
Blackburnian warbler, *Dendroica fusca*
Black-capped chickadee, *Parus atricapilla*
Black-crowned night-heron, *Nycticorax nycticorax*
Black-headed grosbeak, *Pheucticus melanocephalus*
Black-necked stilt, *Himantopus mexicanus*
Blackpoll warbler, *Dendroica striata*
Black tern, *Chlidonias niger*
Black-throated green warbler, *Dendroica virens*
Blue grosbeak, *Guiraca caerulea*
Blue jay, *Cyanocitta cristata*
Blue-winged teal, *Anas discors*
Bobolink, *Dolichonyx oryzivorus*
Bonaparte's gull, *Larus philadelphia*
Brewer's blackbird, *Euphagus cyanocephalus*
Brewer's sparrow, *Spizella breweri*
Broad-winged hawk, *Buteo platypterus*
Brown creeper, *Certhia americana*
Brown-headed cowbird, *Molothrus ater*
Brown thrasher, *Toxostoma rufum*
Bufflehead, *Bucephala albeola*
Bullock's oriole, *see* northern oriole
Burrowing owl, *Athene cunicularia*

California gull, *Larus californicus*
Canada goose, *Branta canadensis*
Canada warbler, *Wilsonia canadensis*
Canvasback, *Aythya valisineria*
Caspian tern, *Sterna caspia*
Cassin's kingbird, *Tyrannus vociferans*
Cassin's sparrow, *Aimophila cassinii*
Cattle egret, *Bubulcus ibis*
Cedar waxwing, *Bombycilla cedrorum*
Cerulean warbler, *Dendroica cerulea*
Chestnut-collared longspur, *Calcarius ornatus*
Chimney swift, *Chaetura pelagica*
Chipping sparrow, *Spizella passerina*
Cinnamon teal, *Anas cyanoptera*
Clark's grebe, *Aechmophorus clarkii*
Clay-colored sparrow, *Spizella pallida*
Cliff swallow, *Hirundo pyrrhonota*
Common barn-owl, *Tyto alba*
Common flicker, *see* northern flicker
Common goldeneye, *Bucephala clangula*
Common grackle, *Quiscalus quiscula*
Common loon, *Gavia immer*
Common merganser, *Mergus merganser*
Common moorhen, *Gallinula chloropus*
Common nighthawk, *Chordeiles minor*
Common poorwill, *Phalaenoptilus nuttallii*
Common redpoll, *Carduelis flammea*
Common snipe, *Gallinago gallinago*
Common tern, *Sterna hirundo*
Common yellowthroat, *Geothlypis trichas*
Connecticut warbler, *Oporornis agilis*
Cooper's hawk, *Accipiter cooperii*
Dark-eyed junco, *Junco hyemalis*
Dickcissel, *Spiza americana*
Double-crested cormorant, *Phalacrocorax auritus*
Downy woodpecker, *Picoides pubescens*
Dunlin, *Calidris alpina*
Eared grebe, *Podiceps nigricollis*
Eastern bluebird, *Siala sialis*
Eastern kingbird, *Tyrannus tyrannus*
Eastern meadowlark, *Sturnella magna*
Eastern phoebe, *Sayornis phoebe*
Eastern screech-owl, *Otus asio*
Eastern wood-pewee, *Contopus virens*

Evening grosbeak, *Coccothraustes vespertinus*
Ferruginous hawk, *Buteo regalis*
Field sparrow, *Spizella pusilla*
Forster's tern, *Sterna forsteri*
Fox sparrow, *Passerella iliaca*
Franklin's gull, *Larus pipixcan*
Gadwall, *Anas strepera*
Golden-crowned kinglet, *Regulus satrapa*
Golden eagle, *Aquila chrysaetos*
Grasshopper sparrow, *Ammodramus savannarum*
Gray catbird, *Dumetella carolinensis*
Gray-cheeked thrush, *Catharus minimus*
Gray partridge, *Perdix perdix*
Great blue heron, *Ardea herodias*
Great crested flycatcher, *Myiarchus crinitus*
Great egret, *Casmerodius albus*
Greater prairie-chicken, *Tympanuchus cupido*
Greater yellowlegs, *Tringa melanoleuca*
Greater white-fronted goose, *Anser albifrons*
Great horned owl, *Bubo virginianus*
Green-backed heron, *Butorides striatus*
Green-winged teal, *Anas crecca*
Gyrfalcon, *Falco rusticolus*
Hairy woodpecker, *Picoides villosus*
Harris' sparrow, *Zonotrichia querula*
Henslow's sparrow, *Passerherbulus henslowii*
Hermit thrush, *Catharus guttatus*
Herring gull, *Larus argentatus*
Hooded merganser, *Lophodytes cucullatus*
Horned grebe, *Podiceps auritus*
Horned lark, *Eremophila alpestris*
House finch, *Carpodacus mexicanus*
House sparrow, *Passer domesticus*
House wren, *Troglodytes aedon*
Hudsonian godwit, *Limosa haemastica*
Indigo bunting, *Passerina cyanea*
Kentucky warbler, *Oporornis formosus*
Killdeer, *Charadrius vociferus*
Lapland longspur, *Calcarius lapponicus*
Lark bunting, *Calamospiza melanocorys*
Lark sparrow, *Chondestes grammacus*
Lazuli bunting, *Passerina amoena*
Least bittern, *Ixobrychus exilis*
Least flycatcher, *Empidonax minimus*

Least sandpiper, *Calidris minutilla*
Least tern, *Sterna antillarum*
LeConte's sparrow, *Ammodramus leconteii*
Lesser golden-plover, *Pluvialis dominica*
Lesser scaup, *Aythya affinis*
Lesser yellowlegs, *Tringa flavipes*
Lincoln's sparrow, *Melospiza lincolni*
Little blue heron, *Egretta caerulea*
Loggerhead shrike, *Lanius ludovicianus*
Long-billed curlew, *Numenius americanus*
Long-billed dowitcher, *Limnodromus scolopaceus*
Long-billed marsh wren, *see* marsh wren
Long-eared owl, *Asio otus*
McCown's longspur, *Calcarius mccownii*
Macgillivray's warbler, *Oporornis tolmiei*
Magnolia warbler, *Dendroica magnolia*
Mallard, *Anas platyrhynchos*
Marbled godwit, *Limosa fedoa*
Marsh wren, *Cistothorus palustris*
Merlin, *Falco columbarius*
Mountain bluebird, *Siala currucoides*
Mountain plover, *Charadrius montanus*
Mourning dove, *Zenaida macroura*
Mourning warbler, *Oporornis philadelphia*
Nashville warbler, *Vermivora ruficapilla*
Northern bobwhite, *Colinus virginianus*
Northern cardinal, *Cardinalis cardinalis*
Northern flicker, *Colaptes auratus* (including "yellow-shafted" *C. a. auratus* and "red-shafted" *C. a. cafer* races)
Northern goshawk, *Accipiter gentilis*
Northern harrier, *Circus cyaneus*
Northern mockingbird, *Mimus polyglottos*
Northern oriole *Icterus galbula* (including "Baltimore," *I. g. galbula* and "Bullock's" *I. g. bullockii* races)
Northern parula, *Parula americana*
Northern phalarope, *see* red-necked phalarope
Northern pintail, *Anas acuta*
Northern rough-winged swallow, *Stelgidopteryx serripennis*
Northern saw-whet owl, *Aegolius acadicus*
Northern shoveler, *Anas clypeata*
Northern shrike, *Lanius excubitor*
Northern waterthrush, *Seiurus novaeboracensis*
Olive-sided flycatcher, *Contopus borealis*
Orange-crowned warbler, *Vermivora celata*

Orchard oriole, *Icterus spurius*
Osprey, *Pandion haliaetus*
Ovenbird, *Seiurus aurocapillus*
Palm warbler, *Dendroica palmarum*
Peregrine falcon, *Falco peregrinus*
Philadelphia vireo, *Vireo philadelphicus*
Pied-billed grebe, *Podilymbus podiceps*
Pine siskin, *Carduelis pinus*
Pinyon jay, *Gymorhinus cyanocephalus*
Piping plover, *Charadrius melodus*
Prairie falcon, *Falco mexicanus*
Purple finch, *Carpodacus purpureus*
Purple martin, *Progne subis*
Pygmy nuthatch, *Sitta pygmaea*
Red-bellied woodpecker, *Melanerpes carolinensis*
Red-breasted nuthatch, *Sitta canadensis*
Red crossbill, *Loxia curvirostra*
Red-eyed vireo, *Vireo olivaceus*
Redhead, *Aythya americana*
Red-headed woodpecker, *Melanerpes erythrocephalus*
Red knot, *Calidris canutus*
Red-necked grebe, *Podiceps grisegena*
Red-necked phalarope, *Phalaropus lobatus*
Red-shafted flicker, *see* northern flicker
Red-tailed hawk, *Buteo jamaicensis*
Red-winged blackbird, *Agelaius phoenicus*
Ring-billed gull, *Larus delawarensis*
Ring-necked duck, *Aythya collaris*
Ring-necked pheasant, *Phasianus colchicus*
Rock dove, *Columba livia*
Rock wren, *Salpinctes obsoletus*
Rose-breasted grosbeak, *Pheucticus ludovicianus*
Ross' goose, *Chen rossii*
Rough-legged hawk, *Buteo lagopus*
Ruby-crowned kinglet, *Regulus calendula*
Ruby-throated hummingbird, *Archilochus colubris*
Ruddy duck, *Oxyura jamaicensis*
Rufous-sided towhee, *Pipilo erythrophthalmus*
Rusty blackbird, *Euphagus carolinus*
Sanderling, *Calidris alba*
Sandhill crane, *Grus canadensis*
Savannah sparrow, *Passerculus sandwichensis*
Say's phoebe, *Sayornis saya*
Scarlet tanager, *Piranga olivacea*

Sedge wren, *Cistothorus platensis*
Semipalmated plover, *Charadrius semipalmatus*
Semipalmated sandpiper, *Calidris pusilla*
Sharp-shinned hawk, *Accipiter striatus*
Sharp-tailed grouse, *Tympanuchus phasianellus*
Sharp-tailed sparrow, *Ammodramus caudacuta*
Short-billed dowitcher, *Limnodromus griseus*
Short-billed marsh wren, *see* sedge wren
Short-eared owl, *Asio flammeus*
Snow bunting, *Plectrophenax nivalis*
Snow goose, *Chen caerulescens*
Snowy egret, *Egretta thula*
Snowy owl, *Nyctea scandiaca*
Solitary sandpiper, *Tringa solitaria*
Solitary vireo, *Vireo solitarius*
Song sparrow, *Melospiza melodea*
Sora, *Porzana carolina*
Spotted sandpiper, *Actitus macularia*
Sprague's pipit, *Anthus spraguei*
Stilt sandpiper, *Calidris himantopus*
Swainson's hawk, *Buteo swainsoni*
Swainson's thrush, *Catharus ustulatus*
Swamp sparrow, *Melospiza georgiana*
Tennessee warbler, *Vermivora peregrina*
Townsend's solitaire, *Myadestes townsendi*
Tree swallow, *Tachycineta bicolor*
Trumpeter swan, *Cygnus buccinator*
Tundra swan, *Cygnus columbianus*
Turkey vulture, *Cathartes aura*
Upland sandpiper, *Bartramia longicauda*
Veery, *Catharus fuscescens*
Vesper sparrow, *Poecetes grammacus*
Violet-green swallow, *Tachycineta thalassina*
Virginia rail, *Rallus limicola*
Warbling vireo, *Vireo gilvus*
Water pipit, *Anthus spinoletta*
Western flycatcher, *Empidonax difficilis*
Western grebe, *Aechmophorus occidentalis*
Western kingbird, *Tyrannus verticalis*
Western meadowlark, *Sturnella neglecta*
Western sandpiper, *Calidris mauri*
Western tanager, *Piranga ludoviciana*
Western wood-pewee, *Contopus sordidulus*

Whimbrel, *Numenius phaeopus*
Whip-poor-will, *Caprimulgus vociferus*
Whistling swan, *see* tundra swan
White-breasted nuthatch, *Sitta carolinensis*
White-crowned sparrow, *Zonotrichia leucophrys*
White-faced ibis, *Plegadis chihi*
White pelican, *see* American white pelican
White-rumped sandpiper, *Calidris fuscicollis*
White-throated sparrow, *Zonotrichia albicollis*
White-throated swift, *Aeronautes saxatilis*
Whooping crane, *Grus americana*
Wild turkey, *Meleagris gallopavo*
Willet, *Catoptrophorus semipalmatus*
Willow flycatcher, *Empidonax traillii*
Wilson's phalarope, *Phalaropus tricolor*
Wilson's warbler, *Wilsonia pusilla*
Winter wren, *Troglodytes troglodytes*
Wood duck, *Aix sponsa*
Wood thrush, *Hylocichla mustelina*
Yellow-bellied flycatcher, *Empidonax flaviventris*
Yellow-bellied sapsucker, *Sphyrapicus varius*
Yellow-billed cuckoo, *Coccyzus americanus*
Yellow-breasted chat, *Icteria virens*
Yellow-headed blackbird, *Xanthocephalus xanthocephalus*
Yellow rail, *Coturnicops noveboracensis*
Yellow-rumped warbler, *Dendroica coronata*
Yellow-shafted flicker, *see* northern flicker
Yellow-throated vireo, *Vireo flavifrons*
Yellow-throated warbler, *Dendroica dominica*
Yellow warbler, *Dendroica petechia*

AMPHIBIANS AND REPTILES (HERPETILES)

Blanding's turtle, *Emydoidea blandingii*
Blue racer, *Coluber constrictor*
Bull frog, *Rana catesbeiana*
Bull snake, *Pituophis catenifer*
Common gartersnake, *Thamnophis sirtalis*
Common racer, *see* blue racer
Common watersnake, *Nerodia sipedon*
Eastern fence lizard, *see* northern prairie lizard
Eastern hognose snake, *Heterodon platyrhinos*
Fence lizard, *see* northern prairie lizard

Great Plains toad, *Bufo cognatus*
Green racer, *see* blue racer
Lesser earless lizard, *Holbrookia masculata*
Many-lined skink, *Eucemes multivirgatus*
Milk snakes, *Lampropeltis triangulum*
Northern cricket frog, *Acris crepitans*
Northern leopard frog, *Rana pipiens*
Northern prairie lizard, *Sceloporus undulatus garmani*
Northern watersnake, *see* common watersnake
Ornate box turtle, *Terrepene ornata*
Painted turtle, *Chrysemys picta*
Plains gartersnake, *Thamnophis radix*
Plains spadefoot toad, *Spea bombifrons*
Prairie-lined racerunner, *see* six-lined racerunner
Prairie rattlesnake, *Crotalus viridis*
Red-sided gartersnake, *see* common gartersnake
Ringneck snake, *Diadophus punctatus*
Rocky Mountain toad, *Bufo woodhousii*
Short-horned horned toad, *Phrynosoma douglassi*
Six-lined racerunner, *Cnemidophorus sexlineatus*
Snapping turtle, *Chelydra serpentina*
Spiny softshell turtle, *Trionyx spiniferus*
Spotted lizard, *see* lesser earless lizard
Striped swift, *see* northern plains lizard
Tiger salamander, *Ambystoma tigrinum*
Wandering gartersnake, *Thamnophis elegans*
Western hog-nosed snake, *Heterodon nasicus*
Western rattlesnake, *see* priairie rattlesnake
Western striped chorus frog, *Pseudacris triseriata*
Woodhouse's toad, *see* Rocky Mountain toad
Yellow mud turtle, *Kinsternon flavescens*

FISH

Bigmouth buffalo, *Ictiobus cyprinellus*
Bigmouth shiner, *Notropis dorsalis*
Black bullhead, *Ictalurus melas*
Black crappie, *Pomoxis nigromaculatus*
Blacknose shiner, *Notropis heterolepis*
Bluegill, *Lepomis macrochirus*
Brassy minnow, *Hybognathus hankinsoni*
Brook stickleback, *Culaea inconstans*
Central stoneroller, *Camptostoma anomalum*

Channel catfish, *Ictalurus punctatus*
Creek chub, *Semotilus atromaculatus*
Emerald shiner, *Notropis atherinoides*
Fathead minnow, *Pimephales promelas*
Finescale dace, *Phoxinus neogaeus*
Flathead catfish, *Pylodictus olivaris*
Flathead chub, *Hybopsis gracilis*
Freshwater drum, *Aplodinotus grunniens*
Gizzard shad, *Dorosoma cepedianum*
Golden shiner, *Notemigonus crysoleucas*
Goldeye, *Hiodon alosoides*
Grass pickerel, *Esox americanus*
Green sunfish, *Lepomis cyanellus*
Iowa darter, *Etheostoma exile*
Longnose dace, *Rhichthys cataractae*
Longnose sucker, *Catostomus catostomus*
Northern pike, *Esox lucius*
Northern redbelly dace, *Phoxinus eos*
Northern redhorse, *see* shorthead redhorse
Orange-spotted sunfish, *Lepomis humulis*
Pearl dace, *Semotilus margarita*
Plains killifish, *Fundulus zebrinus*
Plains minnow, *Hybognathus placitus*
Plains topminnow, *Fundulus sciadicus*
Quillback, *Carpoides cyprinus*
Red shiner, *Notropus lutrensis*
River carpsucker, *Carpoides carpio*
River shiner, *Notropis blennius*
Rock bass, *Ambloplites rupestris*
Sand shiner, *Notropis stramineus*
Sauger, *Stizostedion canadense*
Shorthead redhorse, *Moxostoma macrolepidotum*
Silver chub, *Hybopsis storeriana*
Speckled chub, *Hybopsis aestivalis*
Stonecat, *Noturus flavus*
Suckermouth minnow, *Phenacobius mirabilis*
Tadpole madtom, *Noturus gyrinus*
Topeka Shiner, *Notropus topeka*
Walleye, *Stizostedion vitreum*
White bass, *Morone chrysops*
White sucker, *Catostomus commersoni*
Yellow bullhead, *Ictalurus natalis*
Yellow perch, *Perca flavescens*

INVERTEBRATES

Amphipods (Amphipoda, Crustacea), *Gyraulus, Helisoma, Hyalella, Lymnaea, Physa,* and *Promenetus* spp.

Anostracan shrimp (Anostaca, Branchipoda), *Artemia, Branchinecta, Chiro-cephalopsis, Cyzicus, Eubranchipus,* and *Streptocephalus* spp. *See also* brine shrimp and fairy shrimp

Bogus yucca moths (Proxodidae, Lepidoptera), *Prodoxus* spp.

Brine fly (Ephydriidae, Diptera), *Ephydra hians*

Brine shrimp (Anostraca, Branchipoda), *Artemia salina*

Cladocerans (Cladocera, Branchipoda), *Alona, Bosmina, Ceriodaphnia, Daph-nia, Moina, Pleuoxus,* and *Simocephalus* spp.

Copepods (Copepoda, Branchipoda), esp. *Cyclops, Diaptomus,* and *Cantho-campus* spp.

Dung beetles (Scarabeidae, Coleoptera), Scarabeinae, esp. *Bolboceras* and *Poly-phylla* spp.

Fairy shrimp (Anostraca, Branchipoda), *Branchionecta,* esp. *campestris, lindahli,* and *machini. See also* anostracan shrimp.

Grasshoppers (Acrididae and Tettigoniidae, Orthoptera), mostly gompho-cerine, melanopline, and oedopodine acridids, including Mermira grass-hopper (*Mermira bivitatta*), Haldeman's grasshopper (*Pardalophora hart-manii*), and white-whiskered grasshopper (*Ageneottetix deorum*)

Midges (Chironomidae, Diptera), several genera of Chironominae

Phyllopod shrimp. *See* anostracan shrimp.

Robber flies (Asilidae, Diptera), *Proctacanthus milberti*

Rotifers (Rotifera; Aschelminthes), *Asplanchna, Brachionus, Filinia, Hexartha, Keratella, Notholoca, Platyias* and *Synchaeta* spp.

Swallow bug (Cimicidae, Hemiptera), *Oecicus vicarius*

Yucca moth (Prodoxidae, Lepidoptera), *Tegeticula (Pronuba) yuccasella*

PLANTS

Algae, esp. *Anabaena, Arthrospira, Aphanizomenon, Cladophora, Cosmarium, Micocystis, Nostoc, Oscillatoria,* and *Spirogyra* spp.

Alkali cordgrass, *Spartina gracilis*

American bugleweed, *Lycopus americanus*

American elm, *Ulmus americanus*

American germander, *Teucrium canadense*

American plum, *Prunus americana*

American sloughgrass, *Beckmannia syzigachne*

Annual eriogonum, *Eriogonum annuum*

Arkansas rose, *Rosa arkansana*

Arrow grass, *Triglochin maritima*

Arrowheads, *Sagittaria* spp., esp. *cuneata* and *latifolia*

Balsam poplar, *Populus balsamifera*
Barnyard grass, *Echinochloa crusgalli*
Basswood, *Tilia americana*
Bearded spangletop, *Leptochloa fascicularis*
Beardtongues, *Penstemon* spp., esp. *albidus, angustifolius gracilis, grandiflorus,* and *haydenii*
Bedstraw, *Galium* spp., esp. *aparine, trifidum,* and *trifolium*
Beebalm, *Monarda* spp., esp. *pectinata*
Beggar-ticks, *Bidens* spp., esp. *cernua, comosa, coronata,* and *frondosa*
Big bluestem, *Andropogon gerardii*
Big-tooth aspen, *Populus grandidentata*
Black walnut, *Juglans nigra*
Bladderpod, *Lesquerella ludoviciana*
Bladderworts, *Utricularia* spp.
Blowout grass, *Redfieldia flexuosa*
Blue cardinal-flower, *Lobelia syphilitica*
Blue grama, *Bouteloua gracilis*
Bluejoint, *Calamogrostis canadensis*
Blue skullcap, *Scutellaria lateriflora*
Blue vervain, *Verbena hastata*
Bottlebrush sedge, *Carex hystericina*
Boxelder, *Acer negundo*
Broadleaf milkweed, *Asclepias latifolia*
Brome grasses, *Bromus* spp., esp. *inermis* and *tectorum*
Buckbean, *Menyanthes trifoliata*
Buckthorn, *Rhamnus lanceolata*
Buffaloberry, *Shepherdia argentea*
Buffalo bur, *Solanum rostratum*
Buffalo grass, *Buchloë dactyloides*
Bulrushes, *Scirpus* spp., esp. *acutus, americanus, atrovirens, fluviatilis, paludosus,* and *validus*
Bur oak, *Quercus macrocarpa*
Bur-reed, *Sparganium eurycarpum*
Bush morning-glory, *Ipomoea leptophylla*
Canada wild rye, *Elymus canadensis*
Carolina anemone, *Anemone caroliniana*
Cattails, *Typha latifolia* and *T. angustifolia*
Chair-maker's rush, *see* three-square
Cheatgrass, *Bromus tectorum*
Chokecherry, *Prunus virginiana*
Clammy-weed, *Polanisia dodecandra trachysperma*
Closed gentian, *Gentiana andrewsii*
Collomia, *Collomia linearis*
Common hemicarpha, *Hemicarpha micrantha*

Common mallow, *Malva rotundifolia*

Common reed, *Phragmites australis*

Coneflowers, *Ratibida* spp., esp. *columnifera;* and *Rudbeckia* spp., esp. *hirta*

Coontail, *Ceratophyllum demersum*

Cordgrasses, *Spartina gracilis* and *S. pectinata*

Cottongrass, *Eriophorum polystachion*

Cowlily, *see* yellow water lily

Creeping plox, *Phlox andicola*

Crested shieldfern, *Dryopteris cristata*

Cristatella, *Polanisia jamesii*

Croton, *Croton texensis*

Crowfoots, *Ranunculus* spp., esp. *longirostris* and *scleratus*

Cutleaf ironplant, *Haplopappus spinulosus*

Dayflower, *Commelina virginica*

Dotted gayfeather, *Liatris punctata*

Duckweeds, *Lemna* spp., esp. *minor, perpusilla,* and *trisulca; Wolffia columbiana;* and *Spirodela polyrhiza*

Dwarf sage, *Artemisia cana*

Eastern cottonwood, *Populus deltoides*

Eastern redcedar, *Juniperus virginiana*

Elderberry, *Sambucus canadensis*

False indigo, *Amorpha fruticosa*

Fetid marigold, *Dyssodia papposa*

Field mint, *Mentha arvensis*

Field snake-cotton, *Froelichia floridana*

Fimbristylis, *Fimbristylis* spp.

Flatsedges, *see* umbrella grasses

Floating azollas, *Azolla* spp., esp. *mexicana*

Fourpoint evening primrose, *Oenothera rhombipetala*

Foxtail barley, *Hordeum jubatum*

Gayfeathers, *Liatris* spp., esp. *aspersa, glabrata,* and *punctata*

Giant duckweed, *Spirodella polyrhiza*

Gilia, *Ipomopsis longiflora*

Golden aster, *Crysopsis villosa*

Golden dock, *Rumex maritimus*

Goldenrods, *Solidago* spp., esp. *canadensis, gigantea, missouriensis,* and *rigida*

Goosefoots, *Chenopodium* spp., esp. *album, dissicatum,* and *subglabrum*

Grapes, *Vitis* spp., esp. *riparia* and *vulpina*

Green ash, *Fraxinus pennsylvanica*

Green milkweed, *Asclepias viridiflora*

Hackberry, *Celtis occidentalis*

Hairy four-o'clock, *Mirabilis hirsuta*

Hairy grama, *Bouteloua hirsuta*

Hairy puccoon, *Lithospermum carolinense*

Hardstem bulrush, *Scirpus acutus*
Hayden's penstemon, *Penstemon haydenii*
Hedge nettle, *Stachys palustris*
Hemp, *Cannabis sativa*
Hoary vervain, *Verbena stricta*
Honeylocust, *Gleditsia triacanthos*
Horned pondweed, *Zannichellia palustris*
Horseweed, *Conyza canadensis*
Indiangrass, *Sorghastrum avenaceum*
Ironwood, *Ostrya virginiana*
Juneberry, *see* serviceberry
Junegrass, *Koeleria cristata*
Large beardtongue, *Penstemon grandiflorus*
Leadplant, *Amorpha canescens*
Lemon scurfpea, *Psoralea lanceolata*
Little bluestem, *Andropogon scoparius*
Locoweeds, *Oxytropis* spp., esp. *lambertii*
Loosestrife, *Lythrum dacotanum*
Mannagrasses, *Glyceria* spp., esp. *striata*
Marsh bellflower, *Campanula aparinoides*
Marsh fern, *Thelypteris palustris*
Marsh marigold, *Caltha palustris*
Marsh skullcap, *Scutellaria galericulata*
Mentzelias, *Mentzelia* spp., esp. *nuda* and *decapetala*
Milfoils, *Myriophyllum* spp., esp. *spicatum*
Milkvetches, *Astragalus* spp., esp. *adsurgens, crassicarpus, lotiflorus,* and *missouriensis*
Milkweeds, *Asclepias* spp., esp. *arenaria, incarnata, pumila, stenophylla, syriaca, verticillata,* and *viridiflora*
Miner's candle, *Cryptantha celosioides*
Monkeyflowers, *Mimulus* spp., esp. *glabratus* and *ringens*
Mountain birch, *Betula occidentalis*
Mountain mahogany, *Cercocarpus montanus*
Muskgrass, *Chara* spp., esp. *coronata, fragilis, verrucosa,* and *vulgaris*
Naiads, *Najas* spp., esp. *flexilis* and *guadalupensis*
Narrow beardtongue, *Penstemon angustifolius*
Narrow-leaved puccoon, *Lithosperma incisum*
Needle-and-thread, *Stipa comata*
Needlegrasses, *Stipa comata, S. spartea,* and *S. viridula*
New Jersey tea, *Ceanothus herbaceus*
Nippleweed, *Thelesperma megapotamicum*
Northern bog violet, *Viola nephrophylla*
Northern reedgrass, *Calamogrostris stricta*
Nuttall's sunflower, *Helianthus nuttallii*

Painted milkvetch, *Astragalus ceramicus*
Pale-spike lobelia, *Lobelia spicata*
Paper birch, *Betula papyrifera*
Peachleaf willow, *Salix amygdaloides*
Penstemons, *see* beardtongues
Phragmites, *see* common reed
Pigweeds, *Amaranthus,* spp., esp. *arenicola* and *retroflexus*
Pincushion cactus, *Coryphantha vivipara*
Plains phlox, *Phlox andicola*
Plains sunflower, *Helianthus petiolaris*
Plains yellow primrose, *Calylophus serrulatus*
Platte lupine, *Lupinus plattensis*
Platte thistle, *Cirsium canescens*
Poison ivies, *Toxicodendron radicans* and *T. rydbergii*
Ponderosa pine, *Pinus ponderosa*
Pondweeds, *Potamogeton* spp., esp. *foliosus, gramineus, natans, nodosus, pec-
tinatus, pusillus,* and *richardsonii*
Porcupine grass, *Stipa spartea*
Prairie coneflower, *Ratibida columnifera*
Prairie cordgrass, *Spartina pectinata*
Prairie goldenrod, *Solidago missouriensis*
Prairie larkspur, *Delphinium virescens*
Prairie sage, *Artemisia ludoviciana gnaphalodes*
Prairie sandreed grass, *Calamovilfa longifolia*
Prairie wedgegrass, *Sphenopholis obtusata*
Prairie wild rose, *Rosa arkansana*
Prairie willow, *Salix humilis*
Prickly ash, *Zanthoxylum americanum*
Prickly-pear cacti, *Opuntia* spp., esp. *fragilis* and *macrorhiza*
Prickly poppy, *Argemone polyanthemos*
Purple lovegrass, *Eragrostis spectabilis*
Purple prairie-clover, *Petalostemon purpureum*
Purple three-awn, *Aristida purpurea*
Quaking aspen, *Populus tremuloides*
Ragweeds, *Ambrosia* spp., esp. *artimisiifolia, psilostachya,* and *trifida*
Red ash, *see* green ash
Red false mallow, *see* scarlet mallow
Red three-awn, *Aristida longiseta*
Red-osier dogwood, *Cornus stolonifera*
Red-top, *Agrostis stolonifera*
Reed-canary grass, *Phalaris arundinacea*
Ricegrass, *Oryzopsis hymenoides*
Rocky Mountain beeplant, *Cleome serrulata*
Rocky Mountain juniper, *Juniperus spulorum*

Rocky Mountain pussy-toes, *Antennaria parviflora*
Rough bugleweed, *Lycopus asper*
Rush aster, *Aster junciformis*
Rushes, *Juncus* spp., esp. *balticus, dudleyi, interior, marginatus, nodosus, tenuis,* and *torreyi*
Russian thistle, *Salsola iberica*
Sages, *Artemisia* spp., esp. *campestris, filifolia, frigida,* and *ludoviciana*
Sago pondweed, *Potamogeton pectinatus*
Saltgrass, *Distichlis spicata*
Sandbar willow, *Salix exigua*
Sand bluestem, *Andropogon hallii*
Sand cherry, *Prunus besseyi*
Sand dock, *Rumex venosus*
Sand dropseed, *Sporobolus cryptandrus*
Sand lovegrass, *Eragrostis trichoides*
Sand milkweed, *Asclepias arenaria*
Sand muhly, *Muhlenbergia pungens*
Sand psoralea, *see* lemon scurfpea
Sand sagebrush, *Artemisia filifolia*
Sartwell's sedge, *Carex sartwellii*
Scarlet mallow, *Sphaeralcea coccinea*
Scurfpeas, *Psoralea* spp., esp. *argophylla, digitata, esculenta, lanceolata,* and *tenuiflora*
Sedges, *Carex* spp., esp. *brevior, eleocharis, heliophila, hystericina, lanuginosa, meadii, nebraskensis, praegracilis, sartwellii, scoparia, stipata, stricta,* and *vulpinoides*
Sensitive fern, *Onoclea sensibilis*
Serviceberry, *Amelanchier alnifolia*
Shinners, *Ceanothus ovatus*
Shore buttercup, *Ranunculus cymbalaria*
Showy partridgepea, *Cassia chamaecrista*
Showy vetchling, *Lathyrus polymorphus*
Sideoats grama, *Bouteloua curtipendula*
Silky prairie-clover, *Petalostemon villosum*
Silver-leaf scurfpea, *Psoralea argophylla*
Silver maple, *Acer saccharinum*
Sixweeks fescue, *Festuca octoflora*
Skeletonweed, *Lygodesmia juncea*
Skunkbush sumac, *Rhus aromatica*
Small soapweed, *Yucca glauca*
Smartweeds, *Polygonum* spp., esp. *arenastrum, coccineum, convolvulus, lapathifolium, punctatum,* and *ramosissimum*
Smooth sumac, *Rhus glabra*
Sneezeweed, *Helenium autumnale*

Snowberry, *Symphoricarpos occidentalis*

Speargrass, *see* needle-and-thread

Spiderwort, *Tradescantia occidentalis*

Spikerushes (or spikesedges), *Eleocharis* spp., esp. *acicularis, compressa, eryth-ropoda, macrostachya,* and *smallii*

Spotted touch-me-not, *Impatiens biflora*

Spurges, *Euphorbia* spp., esp. *dentata, geyeri, hexagona, marginata,* and *missurica*

Stiff sunflower, *Helianthus rigidus*

Stoneworts, *see* muskgrass

Sunflowers, *Helianthus* spp., esp. *annuus, maximilianii, nuttallii, petiolaris, rigi-dus,* and *tuberosus*

Swamp lousewort, *Pedicularis lanceolata*

Swamp milkweed, *Asclepias incarnata*

Sweet clover, *Melilotus albus* and *M. officinalis*

Switchgrass, *Panicum virgatum*

Three-awn grasses, *Aristida* spp., esp. *basiramea* and *longiseta*

Three-square, *Scirpus americanus*

Ticklegrass, *Agrostis hyemalis*

Tickseed sunflower, *Bidens coronata*

Tufted loosestrife, *Lysimachia thrysiflora*

Tumbleweed, *Amaranthus* spp., esp. *albus* and *graecizans*

Umbrella grasses, *Cyperus* spp., esp. *aristatus, ferruginescens, rivularis, schwei-nitzii,* and *strigosus*

Virginia creeper, *Parthenocissus vitacea*

Wahoo, *Euonymus atropurpureus*

Watercress, *Nasturtium officinale*

Water crowfood, *Ranunculus* spp., esp. *aquatilis, longirostris,* and *subrigidus*

Water hemlocks, *Cicuta bulbifera* and *C. maculata*

Water horsetail, *Equisetum fluviatile*

Waterlily, *Nymphaea odorata* and *N. tuberosa*

Watermeal, *Wolffia columbiana* and *W. punctata*

Water milfoils, *Myriophyllum* spp., esp. *spicatum*

Water parsnip, *Berula erecta*

Water plantains, *Alisma* spp., esp. *plantago-aquatica*

Waterweeds, *Elodea* spp., esp. *canadensis* and *nuttallii*

Western ragweed, *Ambrosia psilostachya*

Western red lily, *Lilium philadelphicum*

Western wallflower, *Erysimum asperum*

Western wheatgrass, *Agropygron smithii*

Western wild rose, *Rosa woodsii*

Wheatgrasses, *Agropyron* spp., esp. *pectiniforme* and *smithii*

White beardtongue, *Penstemon albidus*

Whitegrass, *Leersia oryzoides*

White sage, *Artemisia ludovicianus*

White-stemmed evening primrose, *Oenothera nuttallii*
Wigeon grass, *Ruppia maritima*
Wild alfalfa, *Psoralea tenuiflora*
Wild barley, *Hordeum jubatum* and *H. pusillum*
Wild begonia, *see* sand dock
Wild lettuce, *Lactuca* spp., esp. *oblongifolia* and *serriola*
Wild licorice, *Glycyrrhiza lepidota*
Wild plum, *Prunus americana*
Wild rice, *Zizania aquatica*
Willowherb, *Epilobium* spp., esp. *adenocaulon*
Willows, *Salix* spp., esp. *amygdaloides, exigua,* and *rigida*
Winged pigweed, *Cycloloma atriplicifolia*
Woolly locoweed, *Oxytropis lambertii*
Yellow pine, *see* ponderosa pine
Yellow water lily, *Nuphar luteum*
Yellow wood sorrel, *Oxalis stricta*
Yucca, *see* small soapweed

Glossary

Adaptation. A genetic trait that increases the ability of an individual organism to survive and reproduce within its environment and thus has been favored by natural selection.

Alkalinity. The capacity of the substances dissolved in a fluid to neutralize acid. Alkaline fluids are those having pH levels above 7.0. Lakes and wetlands may be described, relative to their alkalinity, as slightly to strongly alkaline. *See also* salinity.

Allopatric. Descriptive of two taxa that do not come into geographic contact, at least during their breeding periods. *Cf.* sympatric.

Alluvial. Descriptive of deposits (alluvium) produced by moving water; alluvial strata are those produced by rivers and streams. *See also* colluvial; lacustrine; palustrine.

Annual. A plant or other organism that matures, reproduces, and dies within a single year or growing season.

Anther. The pollen-bearing portion of a flower.

Aquifer. A geologic substrate (rock, sand, gravel) that is permeable, saturated, and able to transmit water in significant quantities under hydraulic gradients. Artesian aquifers are those in which pressurized water flows to the surface without being pumped. When several aquifer subunits are treated as a single geographic unit the term "groundwater reservoir" may be used. *See also* Ogallala aquifer.

Arenicolous. Adapted to sand-living.

Association. A specific type of biotic community, usually named for one or more plant species or genera that consistently occur as dominants within that community. Also used by soil scientists to name collective soil types (soil series groups) that are geographically associated, such as the Valentine association. *See also* community; ecosystem.

Barchan. A crescent-shaped dune characterized by points directed away from, rather than toward, the prevailing winds. *Cf.* parabolic dune.

Bedrock. Any consolidated rock; in Nebraska, generally used to mean pre-Miocene rocks.

Benthos. The bottom-dwelling component of an aquatic ecosystem.

Biennial. A plant or other organism that requires two years to complete its life cycle; in plants, flowering and seed production usually occur only during the second year.

Biocide. A poison designed to kill organisms; insecticides, herbicides, fungicides, and rodenticides are all biocides.

Biota. The combined flora and fauna of an area or region.

Blowout. A localized disruption of the vegetational mantle of a sand dune, causing erosive sand movement.

Bog. A wetland characterized by peat accumulation and poor plant nutrient availability, owing to low rates of organic matter breakdown and resulting highly acidic water characteristics. *See also* fen; marsh.

Boreal. Relating to the north, or of northerly origin.

Brook. A flowing-water environment within a natural channel of the land surface, a river tributary typically less than three meters wide; larger streams are called rivers. Terms such as "creek" and "rivulet" are synonyms; the general term "streams" technically includes both brooks and rivers. *See also* river.

Bunchgrasses. Perennial grasses that grow and expand laterally as distinct clumps, rather than forming relatively continuous ground cover. *Cf.* sod-forming grasses.

Cat-step erosion. Surface erosion pattern (often associated with overgrazing on loess soils) that results when miniature landslides down steep slopes form a series of roughly horizontal steps.

Cenozoic era. The interval encompassing the last 65 million years, also popularly called the "age of mammals." *See also* Quaternary period; Tertiary period.

Cladoceran. A member of a mostly freshwater group (Cladocera, Branchiopoda) of small crustaceans, including such familiar types as water fleas (*Daphnia*).

Climax community. A stable aggregation of interacting species that theoretically is no longer undergoing successional change and is unlikely to alter significantly until the climate changes or outside forces are brought to bear.

Coevolution. The evolution of traits in two interacting species to facilitate some resulting adaptation that is sometimes, but not always, mutually beneficial.

Colluvial. Descriptive of slope-deposited materials (colluvium).

Community. An aggregation of interacting plant and animal populations, especially those occupying a particular localized site or habitat (e.g., the Niobrara riverine forest community). Sometimes used in a more general or abstract sense, such as the "North American desert" or "sand dune" community. *See also* ecosystem; habitat; niche.

Complex dune. A sand dune that comprises two different basic dune types. *See also* compound dune.

Compound dune. A sand dune that comprises two or more dunes of the same general type. *See also* complex dune.

Cool-season grasses. Grasses physiologically adapted for growing in cooler climates, and requiring more water than warm-season grasses; also called C-3 grasses, because of the intermediate 3-carbon molecule stage present during photosynthesis.

Copepods. A group of crustaceans (Copepoda) common in freshwater and marine planktonic ecosystems, notable for their single medial eye and lack of a carapace.

Cursorial. Descriptive of running and adaptations associated with running.

Domelike dune. A generally circular to elliptical sand dune lacking major slip faces. *See also* slip face.

Dominant. Descriptive of the taxa that exert the strongest ecologic influence (control of energy flow) within a community. Also used in ethology to designate high-ranking individual status in social hierarchies, and in genetics to denote relative allele influence in phenotypic expression.

Ecosystem. A complete biotic (floral and faunal) community, together with its physical (abiotic) environment, limited by flows of energy and cycles of materials.

Ecotone. An ecologic transition zone that physically connects two different biotic communities.

Ectothermic. Descriptive of "cold-blooded" animals that depend on external sources of heat to regulate their body temperature. *See also* endothermic.

Endangered. Descriptive of taxa existing in such small numbers as to be in direct danger of extinction without human intervention. *See also* threatened.

Endemic. Descriptive of taxa that are both native to and limited to a specific locality or region. See also exotics; indigenous; introduced.

Endothermic. Descriptive of "warm-blooded" animals that generate and maintain their internal body temperature largely independent of the environment. *See also* ectothermic.

Eolian (or aeolian). Shaped, carried, or influenced by the wind, as sand dunes and loess soils.

Epoch. A geologic interval (such as the Pleistocene epoch) that is a subdivision of a larger "period" (such as the Quaternary period), which in turn is a subdivision of a still larger geologic "era" (such as the Cenozoic era). *See also* period.

Escarpment. A long cliff formed by faulting. The Pine Ridge and Niobrara Valleys lie between escarpments that have been variably exposed by wind and water erosion.

Eutrophic. Descriptive of wetlands that are well supplied with dissolved major nutrients and thus exhibit high biological (photosynthetic) productivity. Eutrophication is the process by which lakes gradually accumulate nutrients and thus "age" over the course of time.

Exotics. Taxa that have been purposely introduced by humans into an area where they are not native.

Extirpated. Descriptive of a taxon (usually a species or subspecies) that has been locally eliminated from an area or region but still persists elsewhere in its overall range.

Fen. A wetland characterized by a boglike substrate accumulation of organic matter (peat) but, unlike bogs, having good plant nutrient levels because of mineral-rich groundwater inflow that prevents the site from becoming overly acidic. *See also* bog; marsh.

Forb. A collective botanical term for nonwoody (herbaceous) plants other than grasses and sedges: i.e., broad-leaved herbs. *See also* herb; shrub.

Formation. A major ecologic community category, often botanically dominated by a single life form of plants, geographically occurring within a generally similar climatic range, and consisting of natural communities having similar successional patterns. Thus, the Nebraska Sandhills region contains the prairie grass, broadleaf forest, and ponderosa pine formations. Also used by geologists to define specific strata (such as the Ash Hollow formation) that typically constitute part of some larger stratigraphic "group" (such as the Ogallala group) and are associated with a particular geologic age. *See also* association; group.

Fossorial. Descriptive of burrowing and digging adaptations.

Gallinaceous. Descriptive of fowllike birds such as grouse, quail, and pheasants.

Groundwater. Subsurface water located within the limits of the water table. *See also* aquifer; water table.

Group. Geologic stratification composed of two or more formations whose origins are geographically and temporally associated: e.g., the Ash Hollow and Valentine formations are both members of the Ogallala group. *See also* formation.

Guild. A group of animals, perhaps of diverse taxonomy, that exploit their resources—especially their foraging niches—in a similar manner; e.g., a guild might comprise ground-foraging and seed-eating birds, mammals, and insects.

Habitat. The biological and physical surroundings of an organism or species; its normal place of residence. Habitats differ from communities in that they include abiotic aspects (e.g,, topography, substrate, climate) of the environment as well as biotic components. *See also* community; niche.

Half-shrub. A perennial plant with a woody base but otherwise mostly herbaceous aboveground parts. *See also* herb; shrub.

Hectare. A metric measurement (10,000 square meters) equal to about 2.47 acres.

Herb. Any plant with no living parts permanently exposed above the ground throughout the year; herbaceous plants are nonwoody, as distinct from half-shrubs, shrubs, and trees. *See also* forb.

Herpetile. A convenient but taxonomically informal collective term for reptiles and amphibians; herpetology is the scientific study of these groups.

High Plains aquifer. See Ogallala aquifer.

Holocene epoch. The roughly 12,000-year interval (within the Quaternary period) extending from the end of the last (Wisconsinian) glaciation to the present time. *See also* Pleistocene epoch.

Home range. An area occupied and regularly traversed by an individual animal over a usually specified course of time (e.g., daily home range, annual home range) but not necessarily defended against intruders. *Cf.* territory.

Hybrid zone. An area of sympatric overlap and interbreeding between two closely related species or well-defined subspecies. Hybrid "suture zones" are areas where multiple hybrid zones occur within the same general region, such as along the Platte and Niobrara Valleys.

Hydrophytic plants. Plants adapted to living in water or water-saturated soils; also called hydrophytes, as distinct from hygrophytes, which are plants adapted to high humidity.

Indigenous. Descriptive of taxa native but not limited to a particular area or region. *Cf.* endemic.

Interdune valley. An area situated between adjacent dunes and often characterized by relatively deeper soils and better moisture relations than dunes. *See also* sand draw.

Introduced. Descriptive of taxa that have accidentally or purposely been added to an area that lies beyond their original distributional limits. *See also* exotics.

Introgressive hybridization. The exchange and dissemination of genes between two hybridizing populations.

Kincaid Act. A federal act of 1904 that provided for establishing homestead claims of 640 acres—one section, or a square mile (about 2.5 square kilometers, or 260 hectares)—in 37 counties of western Nebraska; the earlier Homestead Act limited claims to a quarter-section (160 acres).

Lacustrine (or *lacustrian*). Descriptive of lakes and lake-bed deposits. *See also* palustrine.

Lake. A wetland habitat typically large enough that at least part of its shoreline lacks emergent plants owing to persistent wave action, deep enough that it develops seasonal temperature stratification and thus lacks continuous substrate (benthos) vegetation. *See also* bog; fen; marsh.

Leaching. The downward gravitational movement of dissolved or water-suspended materials through a soil or other substrate.

Linear (or *longitudinal*) *dunes.* Sand dunes with ridges much longer than they are wide, oriented roughly parallel to prevailing winds. Linear dunes have slip faces on both sides of their long axes. *Cf.* transverse dunes.

Littoral zone. The emergent-plant zone on the shores of marshes and lakes.

Loess (pronounced "luss") *or loessial soils.* Silty soils that have been transported and deposited by wind and show little or no vertical stratification. *See also* eolian.

Marsh. A wetland that is typically too small in area to have significant wave action—thus often permitting a continuous littoral zone of emergent vegetation—and sufficiently shallow that its bottom may be entirely vegetated with aquatic plants. *See also* bog; fen; lake.

Mesic. Descriptive of an intermediate ecological condition between xeric and hydric, especially with regard to soil moisture. Plants adapted to mesic conditions are often called mesophytic or mesophytes. *See also* hydrophytic plants; xeric.

Metheglobinemia. A physiological syndrome associated with reduced oxygen-carrying capability of the blood; it can develop in infants who ingest excessive nitrates.

Miocene epoch. The interval within the Tertiary period (and the associated geologic strata deposited during that interval) from the end of the preceding

Oligocene epoch (24 million years ago) to the start of the Pliocene epoch (5 million years ago). The sands and gravels of the Ogallala group and part of the older (lower) Arikaree group are Miocene strata underlying the Sandhills region. *See also* Pliocene epoch.

Niche. An organism's specific role or biological "profession" within its ecologic environment, defined by its behavioral, morphological, and anatomical adaptations to that environment. *See also* habitat.

Ogallala aquifer. The principal aquifer of the Nebraska Sandhills, comprising Miocene, Pliocene, and Pleistocene deposits of sand and gravel. It includes most or all (depending on definitions) of the High Plains aquifer, which extends from southern South Dakota and eastern Wyoming to northwestern Texas and eastern New Mexico. *See also* aquifer.

Ogallala group. The variably thick (30 to 250 meters) strata of alluvial sands and gravels deposited in Nebraska during the late Miocene epoch. These mostly originated in ancient rivers of Wyoming and Colorado and now constitute much of the Ogallala aquifer. *See also* group.

Paleovalley. A valley of the geological past, often buried under more recent sediments. Paleorivers are similarly extinct rivers.

Palustrine (or palustrian). Descriptive of marshes and marsh-related deposits. *See also* lacustrine.

Pandemic. Descriptive of a taxon having a very widespread distribution.

Parabolic dune. A crescent-shaped dune form whose points are directed toward, rather than away from, the prevailing winds. *Cf.* barchan.

Passerine. Descriptive of members of the avian order Passeriformes, often popularly called "songbirds."

Perennial. A plant of indefinite life span, typically lasting at least three years.

Period. A geologic interval that is shorter than an era but longer than an epoch, such as the Quaternary period. *See also* epoch.

Photoperiod. The relative length of daylight—increasing during spring and diminishing during autumn—within a 24-hour period.

Phyllopods. A group of aquatic crustaceans (Branchiopoda) with flattened, leaf-like legs that are used in respiration as well as for locomotion. Among those common in highly alkaline wetlands of the western Sandhills are brine shrimp and fairy shrimp.

Plankton. Small (often only microscopically visible) unattached aquatic organisms, including chlorophyll-containing forms such as algae (phytoplankton) and invertebrates such as rotifers and microscopic crustaceans (zooplankton).

Playa. A Spanish-language term (meaning "shore" or "beach") that refers to the generally intermittent and often highly alkaline or saline wetlands of the American Southwest. The roughly 2,500 alkaline wetlands of the Sandhills are similar to typical playas in their high annual average alkalinity (pH 7.8–10.8), high sodium and potassium concentrations, and high carbonate and bicarbonate levels; they are very low in chlorides and sulfates, however, and also exhibit marked seasonal variations in their alkalinity and in their algae and invertebrate populations.

Pleistocene epoch. The interval (within the Quaternary period) extending from about 1.6 million years ago (the end of the Tertiary period's Pliocene epoch) to 12,000 years ago. Popularly referred to as the "ice age," this epoch includes four major glacial periods (and their associated interglacial intervals), of which the Nebraskan glaciation was the earliest and the Wisconsinian glaciation the most recent. *See also* Holocene epoch.

Pliocene epoch. The interval (and the associated geologic strata deposited during that interval) extending from the end of the preceding Miocene epoch (about 5 million years ago) to the start of the Pleistocene epoch (about 1.7 million years ago).

Prairie. A native plant community dominated by perennial grasses. Rates of effective precipitation (precipitation-to-evaporation ratios) in prairies are typically too low to support tree-dominated communities but higher than those characteristic of deserts. Prairies may be described by the relative stature of their dominant grasses (shortgrass, mixed-grass, or tallgrass prairies), by the particular dominant grass taxa (e.g., big bluestem or grama grass), or by the characteristic form of these grasses (bunchgrasses vs. sod-forming grasses). Prairie soils are usually high in organic matter, calcium, and other inorganic plant nutrients, but this is not true of the poorly developed Sandhills soils.

Psammophytes. Sand-adapted plants.

Quaternary period. The second subdivision of the Cenozoic era, extending from roughly 1.6 million years ago to the present time. *See also* Tertiary period; Holocene epoch; Pleistocene epoch.

Refugium. A location or region, such as the Niobrara Valley, where certain taxa are able to survive after the environment has altered and caused these organisms to be otherwise eliminated from surrounding areas. *See also* relicts.

Relicts. Plant or animal taxa that persist locally in an otherwise generally altered habitat or climate. *See also* refugium.

Resource partitioning. The subdivision by two or more species of a community's important resources (food, space, vegetative cover) in ways that are adaptively related to these species' differing niches.

Rhizome. An underground stem system that produces new roots and aboveground stems laterally, thus increasing the plant's surface area through a process called "tillering."

Riparian. Relating to the shorelines of a river or lake.

River. A flowing-water habitat in which water moves horizontally within a surface channel that typically has been variably modified by the eroding and depositional actions of the water itself. *See also* brook.

Riverine. Relating to a river. Forests limited to river valleys are often called riverine, riparian, or gallery forests.

Rotifers. A group of microscopically small invertebrates common in fresh water, named for the crown of cilia that gives the impression of a rotating wheel.

Ruderals. Weedy plants, typically early invaders ("pioneers") during ecologic succession.

Salinity. The relative concentration of dissolved salts in a fluid. *See also* alkalinity.

Saltatorial. Descriptive of jumping, and jumping adaptations. Also used to describe the bounding movements of wind-carried sand grains over the windward surfaces of a dune.

Sand draw. A local vernacular term for any interdune valley that is subjected to occasional flash floods and thus is often nearly or entirely vegetation-free.

Sandhills prairie. A type of North American grassland that has developed over highly sandy soils and is dominated by a mixture of short, medium, and tall perennial grasses.

Sand sheets. Sand accumulations having no surface topography of distinctive shape.

Sand sage prairie. An aridity-adapted variant of North American prairie, found locally in southwestern Nebraska in relatively dry regions beyond the Sandhills proper, dominated by sand sage as well as many plant species common to the Sandhills.

Sedge. A general term for members of the family Cyperaceae, which are grasslike plants having solid, triangular stems, rather than the round and hollow stems typical of true grasses (family Poaceae).

Shortgrass prairie. Perennial grasslands of western North America that occur in areas too arid to support mixed-grass prairies. *See also* steppe.

Shrub. A general term for woody plants that are usually under three meters tall at maturity and often have multiple stems. *See also* herb.

Slip face. The leeward and generally steeper side of a dune (inclining typically 30–34 degrees), where loose sand grains that have been carried by wind over the dune's crest slide down its face, causing surface instability and resulting in gradual dune movement.

Sod-forming grasses. Grasses having root systems that tend to bind the ground substrate in a continuous, rather than discontinuous, manner. *Cf.* bunchgrasses.

Soil series. A grouping of similar soils that have closely related soil "profiles" (horizontal layers having various physical and chemical characteristics). *See also* association.

Speciation. The proliferation of species through processes of evolutionary differentiation.

Steppe. A Russian-language term for native shortgrass communities; comparable to the popular usage of "plains" in North America.

Stratigraphy. An area of geology concerned with the sequential ordering and geographic distributions of rock strata.

Succession. A series of gradual ecologic changes in biotic communities over time, as relatively temporary (successional or seral) taxa are sequentially replaced by others that are able to persist and reproduce successfully for a more prolonged or even indefinite period (so-called "climax" species). *See also* climax community.

Sympatric. Descriptive of two species having significant geographic overlap, at least during their breeding season. *Cf.* allopatric.

Tallgrass prairie (or *"true" prairie*). Perennial grasslands that are dominated by tall-stature grasses, often at least two meters high. These typically occur in areas that are more mesic than those supporting mostly lesser-stature grasses (mixed-grass prairies) but are still too arid—or too frequently burned—to support forests.

Taproot. A vertically oriented, variably thickened root.

Taxon (pl. taxa). A particular taxonomic category such as a species or genus or, especially, a representative member of that category.

Territory. In ornithology, an area defended against conspecifics and other inter-acting species, typically dominated by an individual bird (usually the male) or by a pair for a particular period—usually during the breeding season (breeding territories) but occasionally at other times (winter territories or permanent territories). Similar territoriality occurs in some mammals, reptiles, amphibians, and fish. *Cf.* home range.

Tertiary period. The first of the two major subdivisions of the Cenozoic era, beginning about 65 million years ago and lasting until about 1.6 million years ago. *See also* Quaternary period.

Thermal stratification. The layering of a lake's waters, caused by temperature-related variations in water densities.

Threatened (or *vulnerable*). Descriptive of taxa that have declined and exist in small numbers but are not yet believed to be endangered.

Transverse dunes. Sand dunes whose parallel ridges are oriented at approximate right angles to the prevailing winds, with their slip faces on the leeward side (thus resembling waves of water), including barchanoid-ridge and domal-ridge types. *Cf.* linear dunes.

Tuber. An enlarged part of an underground plant stem, such as a potato.

Ungulate. A hooved mammal.

Warm-season grasses. Grasses physiologically adapted to grow in warmer periods having greater water stress; also called C-4 grasses because their photosynthetic pathway involves an intermediate 4-carbon molecule stage. *Cf.* cool-season grasses.

Water table. The zone of a saturated substrate that begins where the water pressure equals the atmospheric pressure, producing the top boundary of the saturated portion of the substrate. Water thus no longer moves downward but instead may move laterally, through an aquifer. *See also* aquifer; groundwater.

Xeric. Descriptive of an ecologic condition of marked aridity, especially with regard to soil. Xerophytic plants ("xerophylls") are those adapted to arid conditions. *See also* mesic.

Bibliography

Adelman, D. D., W. J. Schroeder, R. J. Smaus, and G. P. Wallin. 1985. Overview of nitrate in Nebraska's ground water. *Trans. Neb. Acad. Sci.* 13:1175–81.

Ahlbrandt, T. S., and S. G. Fryberger. 1980. Eolian deposits in the Nebraska Sand Hills. In *Geologic and paleoecologic studies of the Nebraska Sand Hills,* pp. 1–24. U.S. Geol. Surv. Prof. Pap. 1120 [Reston VA: U.S. Dept. Inter., Geol. Survey.]

——, J. B. Swinehart, and D. G. Marony. 1983. The dynamic Holocene dune fields of the Great Plains and Rocky Mountain basins. In *Eolian sediments and processes: Developments in sedimentology,* ed. M. E. Brookfield and T. S. Ahlbrandt, pp. 379–406. Amsterdam: Elsevier.

American Ornithologists Union. 1983. *A.O.U. Checklist of North American Birds.* Washington DC: A.O.U. (6th ed.).

Armstrong, D. M. 1972. *Distribution of mammals in Colorado.* Monog. Mus. Nat. Hist. no. 3. Lawrence: Univ. of Kansas.

——, J. R. Choate, and J. K. Jones Jr. 1986. *Distributional patterns of mammals in the plains states.* Occ. Papers no. 105. Lubbock: Texas Tech. Univ. Mus.

Bailey, V., and C. C. Sperry. 1929. Life history and habits of grasshopper mice, genus *Onychomys. U.S. Dept. Agr. Tech. Bull.* 145:1–19.

Ballinger, R. E., and S. M. Jones. 1985. Ecological disturbance in a Sandhills prairie: Impact and importance to the lizard community on Arapaho Prairie in western Nebraska. *Prairie Nat.* 17:91–100.

——, S. M. Jones, and J. W. Neitfeld. 1990. Patterns of resource use in a lizard community in the Nebraska Sandhills prairie. *Prairie Nat.* 22:75–86.

——, J. D. Lynch, and P. H. Cole. 1979. Distribution and natural history of amphibians and reptiles in western Nebraska, with ecological notes on the herpetiles of Arapaho Prairie. *Prairie Nat.* 11:65–74.

Banks, R. C., R. W. McDiarmid, and A. L. Gardner. 1987. *Checklist of vertebrates of the United States, the U.S. Territories, and Canada.* Res. Publ. no. 166. Washington DC: U.S. Dept. of the Interior.

Barnes, P. W. 1980. Water relations and distributions of several dominant grasses in a Nebraska Sandhills prairie. M.S. thesis, Univ. of Nebraska–Lincoln.

——. 1985. Adaptation to water stress in the big bluestem–sand bluestem complex. *Ecology* 66:1908–20.

——, and A. T. Harrison. 1982. Species distribution and community organization in a Nebraska Sand Hills prairie as influenced by plant/soil water relationships. *Oecologia* 52:192–201.

——, A. T. Harrison, and S. P. Heinisch. 1984. Vegetation patterns in relation to topography and edaphic variation in Nebraska Sand Hills prairie. *Prairie Nat.* 16:145–58.

Bartholomew, G. A., Jr., and H. H. Caswell Jr. 1951. Locomotion in kangaroo rats and its adaptive significance. *J. Mamm.* 32:155–69.

Baumann, W. L. 1982. Microhabitat use in three species of rodents on a Nebraska Sandhills prairie. M.S. thesis, Univ. of Nebraska–Lincoln.

Bee, J. W., G. E. Glass, R. S. Hoffmann, and R. R. Patterson. 1981. *Mammals in Kansas.* Mus. Nat. Hist. Public Ed. Ser. no. 7. Lawrence: Univ. of Kansas.

Beed, W. E. 1936. *A preliminary study of the animal ecology of the Niobrara Game Preserve.* Bull. 10. Lincoln: Univ. Neb. Cons. and Surv. Div.

Bentall, R. 1989. Streams. In Bleed and Flowerday 1989, pp. 93–114.

Bessey, C. E. 1900. Vegetation of the Sandhills. In *Ann. Rept.,* pp. 81–95. Lincoln: Neb. State Board Agr.

Bicek, T. K. 1977. Some eco-ethological aspects of a breeding population of long-billed curlews *(Numenius americanus)* in Nebraska. M.A. thesis, Univ. of Nebraska–Omaha.

Bleed, A. 1989. Groundwater. In Bleed and Flowerday 1989, pp. 67–92.

——, and C. A. Flowerday, eds. 1989. *An atlas of the Sand Hills.* Resource Atlas No. 5. Lincoln: Univ. Neb. Cons. and Surv. Div. (Rev. ed., 1993.)

——, and M. Ginsberg. 1989. Lakes and wetlands. In Bleed and Flowerday 1989, pp. 115–22.

Bomberger, M. L. 1983. The breeding and ecology of the Wilson's phalarope in the Nebraska Sandhills. M.S. thesis, Univ. of Nebraska–Lincoln.

——, S. L. Shields, A. T. Harrison, and K. H. Keeler. 1983. Comparison of old field succession on a tallgrass prairie and a Nebraska Sandhills prairie. *Prairie Nat.* 15:9–15.

Borchert, J. R. 1950. The climate of the central North American grassland. *Ann. Assoc. Am. Geog.* 40:1–39.

Bose, D. R. 1977. *Rangeland resources of Nebraska.* Lincoln: U.S. Soil Cons. Service.

Brogie, M. A., and M. J. Mossman. 1983. Spring and summer birds of the Niobrara Valley Preserve area, Nebraska. *Neb. Bird Rev.* 51:44–51.

Brown, C. R. 1985. The costs and benefits of coloniality in the cliff swallow. Ph.D. diss., Princeton Univ.

——. 1988. Enhanced foraging efficiency through information centers: A benefit of coloniality in cliff swallows. *Ecology* 69:602–13.

——, and M. B. Brown. 1986. Ectoparasitism as a cost of coloniality in cliff swallows *(Hirundo pyrrhonota)*. *Ecology* 67:1206–18.

——, and M. B. Brown. 1988. The costs and benefits of egg destruction by conspecifics in colonial cliff swallows. *Auk* 105:737–48.

——, and M. B. Brown. 1989. How many swallows make a summer? *Nebraskaland* 67 (3): 6–13.

——, and M. B. Brown. 1990. The great egg scramble. *Natural History* 99 (2): 34–41.

——, and M. B. Brown. 1991. Selection of high-quality host nests by parasitic cliff swallows. *Anim. Behav.* 41:457–65.

——, and J. L. Hoogland. 1986. Risk in mobbing for solitary and colonial swallows. *Anim. Behav.* 34:1319–23.

Brown, J. H. 1973. Species diversity of seed-eating desert rodents in sand dune habitats. *Ecology* 54:775–87.

Bruner, L. 1902. A comparison of the bird-life found in a Sandhills region of Holt County in 1883–84 and in 1901. *Proc. Neb. Ornith. Union* 3:58–63.

Buckwalter, D. W. 1983. *Monitoring Nebraska's Sandhills lakes.* Resource Rept. no. 10. Lincoln: Univ. Neb. Cons. and Surv. Div.

Burzlaff, D. F. 1960. Soil as a factor influencing distribution of vegetation in the Sandhills of Nebraska. Ph.D. diss., Univ. of Nebraska–Lincoln.

——. 1962. *A soil and vegetation inventory and analysis of three Nebraska Sandhill range sites.* Neb. Agr. Exp. Sta. Res. Bull. 206. Lincoln: Univ. of Nebraska Coll. of Agr.

Carpenter, C. C. 1960. Aggressive behavior and social dominance in the six-lined racerunner. *Anim. Behav.* 8:61–66.

——. 1962. Patterns of behavior in two Oklahoma lizards [*Sceloporus undulatus* and *Cnemidophorus sexlineatus*]. *Am. Midl. Nat.* 67:132–51.

Clawson, S. D. 1980. Comparative ecology of the northern oriole *(Icterus galbula)* and the orchard oriole *(Icterus spurius)* in western Nebraska. M.S. thesis, Univ. of Nebraska–Lincoln.

Cockrum, E. L. 1952. *Mammals of Kansas.* Mus. Nat. Hist. Publ. no. 7. Lawrence: Univ. of Kansas.

Collins, J. T. 1982. *Amphibians and reptiles in Kansas.* Mus. Nat. Hist. Pub. Ed. Ser. no. 8. Lawrence: Univ. of Kansas.

Condra, G. E. 1908. *The geography of Nebraska.* Lincoln: Univ. Pub.

Conservation and Survey Division. 1986. *The groundwater atlas of Nebraska.* Resource Atlas No. 4. Lincoln: Univ. Neb. Cons. and Surv. Div.

Cox, M. K., and W. L. Franklin. 1989. Terrestrial vertebrates of Scotts Bluff National Monument, Nebraska. *Southwestern Nat.* 49:597–613.

Cramer, T. 1993. Limnological and habitat influences on wetland bird communities of the Nebraska Sandhills. M.S. thesis, Univ. of Nebraska–Lincoln.

Cramp, S., and K. E. L. Simmson, eds. 1977. *The birds of the western Palearctic.* Vol. 1. Oxford: Oxford Univ. Press.

Cross, F. B., and J. T. Collins. 1975. *Fishes in Kansas.* Mus. Nat. Hist. Pub. Ed. Ser. no. 2. Lawrence: Univ. of Kansas.

Currier, P. J., G. R. Lingle, and J. G. VanDerwalker. 1985. *Migratory bird habitat on the Platte and North Platte rivers in Nebraska.* Grand Island NE: Platte River Whooping Crane Critical Habitat Maintenance Trust.

Daley, R. H. 1972. The native sand sage vegetation of eastern Colorado. M.S. thesis, Colorado State Univ.–Fort Collins.

Desmond, M. J. 1991. Ecological aspects of burrowing owl nesting strategies in the Nebraska panhandle. M.S. thesis, Univ. of Nebraska–Lincoln.

Dice, L. R. 1941. Variation of the deer-mouse *(Peromyscus maniculatus)* on the Sand Hills of Nebraska and adjacent areas. *Univ. Mich. Contr. Lab. Vert. Genet.* 15:1–19.

Dort, W., and J. N. Jones, eds. 1970. *Pleistocene and recent environments of the central Great Plains.* Univ. Kansas Dept. of Geology Spec. Pub. no. 3. Lawrence: Univ. Press of Kansas.

Droge, D. L. 1980. Seasonal patterns of aggression in the lizards *Sceloporus undulatus* and *Holbrookia maculata.* M.S. thesis, Univ. of Nebraska–Lincoln.

——, S. M. Jones, and R. E. Ballinger. 1982. Reproduction of *Holbrookia maculata* in western Nebraska. *Copeia* 1982:356–62.

Ducey, J. E. 1984. Location and habitat size of lakes in the Nebraska Sandhills utilized by trumpeter swans. *Neb. Bird Rev.* 52:19–22.

——. 1988. *Nebraska birds: Breeding status and distribution.* Omaha: Simmons-Boardman.

Egoscue, H. J. 1960. Laboratory and field studies of the northern grasshopper mouse. *J. Mamm.* 91:99–110.

Eisenberg, J. F. 1963a. The behavior of heteromyid rodents. *Univ. Calif. Publ. Zool.* 69:1–114.

——. 1963b. A comparative study of sandbathing behavior in heteromyid rodents. *Behaviour* 22:16–23.

Engberg, R. A. 1967. *The nitrate hazard in well water, with special reference to Holt County, Nebraska.* Water Res. Paper 4. Lincoln: Univ. Neb. Cons. and Surv. Div.

——, and R. F. Spalding. 1978. *Groundwater quality atlas of Nebraska.* Resource Atlas No. 3. Lincoln: Univ. Neb. Cons. and Surv. Div.

Environmental Protection Agency. 1990. *National pesticide survey.* Washington DC.

Erickson, N. E., and D. M. Leslie Jr. 1987. *Soil-vegetation correlations in the Sandhills and Rainwater Basin wetlands of Nebraska.* U.S. Fish and Wildlife Serv. Biol. Rept. 87 (11). Washington DC: U.S. Dept. Inter.

Exner, M. E., and R. F. Spalding. 1990. *Occurrence of pesticides and nitrate in Nebraska's ground water.* Lincoln: Water Center, Inst. of Agr. and Nat. Res., Univ. of Nebraska.

Faanes, C. A. 1984. Wooded islands in a sea of prairie. *Am. Birds* 38:3–6.

Farrar, J. 1984. Trumpeters. *Nebraskaland* 62 (2): 22–29.

——. 1990. *Field guide to wildflowers of Nebraska and the Great Plains.* Lincoln: Neb. Game and Parks Comm.

——. 1992. Platte River instream flow: Who needs it? *Nebraskaland* 70 (10): 38–47.

Ferguson, G. W., C. H. Bohlen, and H. P. Woolley. 1980. *Sceloporus undulatus:* Comparative life history and regulation of a Kansas population. *Ecology* 61:313–22.

Fichter, E. 1954. An ecological study of invertebrates of grassland and deciduous shrub savanna in eastern Nebraska. *Am. Midl. Nat.* 51:321–439.

Fitch, H. S. 1958. Natural history of the six-lined racerunner. *Univ. Kansas Mus. Nat. Hist. Pub.* 11:11–62.

——. 1970. *Reproductive cycles in lizards and snakes.* Mus. Nat. Hist. Misc. Publ. no. 52. Lawrence: Univ. of Kansas.

Fitzgerald, F. G., and R. J. Wooton. 1992. Behavioural ecology of sticklebacks. In *Behaviour of teleost fishes,* ed. T. J. Pitcher. London: Chapman & Hall.

Flake, L. D. 1973. Food habits of four species of rodents on a short-grass prairie in Colorado. *J. Mamm.* 54:636–47.

——. 1974. Reproduction of four rodent species in a shortgrass prairie of Colorado. *J. Mamm.* 55:213–16.

Flores, R. M., and S. S. Kaplan, eds. 1985. *Cenozoic paleogeography of the west-central U.S.* Soc. Econ. Paleontol. and Minerol., Rocky Mtn. Sect., Rocky Mtn. Paleogeogr. Symp. 3, Denver.

Flowerday, C. A., ed. 1993. *Flat water: A history of Nebraska and its water.* Resource Rpt. No. 12. Lincoln: Univ. Neb. Cons. and Surv. Div.

Forbes, W. T. M. 1923. *The Lepidoptera of New York and neighboring states.* Ithaca NY: Cornell Univ. Agr. Exp. Sta. Mem. no. 68.

Frank, D. H. 1989. Spatial organization, social behavior, and mating strategies of the southern grasshopper mouse *(Onychomys torridus)* in southeastern Arizona. Ph.D. diss., Cornell Univ.

Freeman, P. 1989a. Amphibians and reptiles. In Bleed and Flowerday 1989, pp. 157–60.

——. 1989b. Mammals. In Bleed and Flowerday 1989, pp. 181–88.

Frolik, A. L., and F. D. Keim. 1933. Native vegetation in the prairie hay district of north-central Nebraska. *Ecology* 14:298–305.

Ginsberg, M. H. 1985. Nebraska's Sandhills lakes: A hydrogeological overview. *Water Resources Bull.* 21:573–78.

Great Plains Flora Assoc. 1977. *Atlas of the flora of the Great Plains.* Ames: Iowa State Univ. Press.

Green, N. E. 1969. Occurrence of small mammals on sandhill rangelands in eastern Colorado. M.S. thesis, Colorado State Univ.–Fort Collins.

Gubanyi, J. A. 1989. Habitat use and diet analysis of barn owls in western Nebraska. M.S. thesis, Univ. of Nebraska–Lincoln.

Gutentag, E. D., F. H. Heimes, N. C. Krothe, R. R. Luckey, and J. B. Weeks. 1984. *Geohydrology of the High Plains aquifer in parts of Colorado, Kansas, Nebraska, New Mexico, Oklahoma, South Dakota, Texas, and Wyoming.* Washington DC: U.S. Geol. Survey.

Haff, P. K. 1986. Booming dunes. *Am. Scientist* 74:376–81.

Hafner, M. S., and D. J. Hafner. 1979. Vocalizations of grasshopper mice (genus *Onychomys*). *J. Mamm.* 60:85–94.

Hansen, C. M., and E. D. Fleharty. 1974. Structural ecological parameters of a population of *Peromyscus maniculatus* in west-central Kansas. *Southwestern Nat.* 19:293–303.

Hardy, D. F. 1962. Ecology and behavior of the six-lined racerunner, *Cnemidophorus sexlineatus. Univ. Kansas Sci. Bull.* 43:3–73.

Haskens, B. 1980. Seed predation and fertility in *Yucca glauca.* B.S. honors thesis, Univ. of Nebraska–Lincoln.

Heinisch, S. P. 1981. Water allocation and rooting morphology of two *Bouteloua* species in relation to their distribution patterns in Nebraska Sandhills. M.S. thesis, Univ. of Nebraska–Lincoln.

Hergenrader, G. L. 1993. The "reforestation" of the Sandhills. *Nebraskaland* 71 (1): 64–65.

Hines, T. 1980. An ecological study of *Vulpes velox* in Nebraska. M.S. thesis, Univ. of Nebraska–Lincoln.

Hiskey, R. M. 1981. The trophic dynamics of an alkaline-saline Nebraska Sandhills lake. Ph.D. diss., Univ. of Nebraska–Lincoln.

Hoesel, S. F. 1973. The impact of center-pivot irrigation on the Sand Hills of Nebraska: Brown County, a case study. M.A. thesis, Univ. of Nebraska–Omaha.

Hoffmeister, D. R. 1986. *Mammals of Arizona.* Tucson: Univ. of Arizona Press and Arizona Game and Fish Dept.

Hopkins, H. H. 1951. Ecology of the native vegetation of the loess hills in central Nebraska. *Ecol. Monog.* 21:125–47.

——. 1952. Native vegetation of the loess hills–Sandhills ecotone in Nebraska. *Trans. Kans. Acad. Sci.* 55:267–77.

Horner, B. E., and J. M. Taylor. 1968. Growth and reproductive behavior in the southern grasshopper mouse. *J. Mamm.* 49:644–60.

Hrabik, R. A. 1989. Fishes. In Bleed and Flowerday 1989, pp. 143–54.

Hudson, G. E. 1942. The amphibians and reptiles of Nebraska. Bull. 24. Lincoln: Univ. Neb. Cons. and Surv. Div.

Hulett, C. K., C. D. Sloan, and G. W. Tomenek. 1968. The vegetation of remnant grasslands in the loessial region of northwestern Kansas and southwestern Nebraska. *Southwestern Nat.* 13:377–91.

Hunt, J. C. 1965. The forest that men made. *Am. Forests Mag.* 71 (11): 19–22, 46–48; 71 (12): 32–34, 48–50.

Imler, R. 1945. Bullsnakes and their control on a Nebraska wildlife refuge. *J. Wildl. Mgmt.* 9:265–73.

Iverson, J. B. 1990. Nesting and parental care in the mud turtle *Kinosternon flavescens. Can. J. Zool.* 68:230–33.

——. 1991. Life history and demography of the yellow mud turtle (*Kinosternon flavescens*). *Herpetologica* 47:371–93.

——, and G. R. Smith. 1993. Reproductive ecology of the painted turtle (*Chrysemys picta*) in the Nebraska Sandhills. *Copeia* 1993:1–21.

Jameson, E. W., Jr. 1947. Natural history of the prairie vole (mammalian genus *Microtus*). *Univ. Kansas Mus. Nat. Hist. Pub.* 1:125–51.

Jensen, R. 1992. *Groundwater in the Great Plains.* Great Plains Agr. Council Rept. no. 141. College Station: Texas Water Resources Institute.

Joern, A. 1982. Distributions, densities, and relative abundances of grasshoppers (Orthoptera: Acrididae) in a Nebraska Sandhills prairie. *Prairie Nat.* 14:37–45.

———. 1988. Foraging behavior and switching by the grasshopper sparrow, *Ammodrammus savannarum:* Searching for multiple prey against a heterogeneous background. *Am. Midl. Nat.* 119:225–34.

———. 1992. Variable impact of avian predation on grasshopper assemblies in Sandhills grassland. *Oikos* 64:458–63.

Johnsgard, P. A. 1978. The ornithogeography of the Great Plains states. *Prairie Nat.* 10:97–112.

———. 1979a. The breeding birds of Nebraska. *Neb. Bird Rev.* 47:3–16.

———. 1979b. *Birds of the Great Plains: Breeding species and their distribution.* Lincoln: Univ. of Nebraska Press.

———. 1984. *The Platte: Channels in time.* Lincoln: Univ. of Nebraska Press.

———. 1986. *A revised list of the birds of Nebraska and adjacent plains states.* Occ. Pap. Neb. Ornith. Union No. 6. (3d corr. and rev. ed.).

Jones, J. K., Jr. 1964. *Distribution and taxonomy of mammals of Nebraska.* Publ. Mus. Nat. Hist. no. 16. Lawrence: Univ. of Kansas.

———, D. M. Armstrong, R. S. Hoffmann, and C. Jones. 1983. *Mammals of the northern Great Plains.* Lincoln: Univ. of Nebraska Press.

———, D. M. Armstrong, and J. R. Choate. 1985. *Guide to mammals of the Plains states.* Lincoln: Univ. of Nebraska Press.

Jones, J. O. 1990. *Where the birds are: A guide to all 50 states and Canada.* New York: Morrow.

Jones, S. M., and R. E. Ballinger. 1987. Comparative life histories of *Holbrookia maculata* and *Sceloporus undulatus* in western Nebraska. *Ecology* 68:1828–38.

———, and D. L. Droge. 1980. Home range size and spatial distribution of two sympatric lizard species *(Sceloporus undulatus, Holbrookia maculata)* in the Sand Hills of Nebraska. *Herpetologica* 36:127–32.

Judd, S. D. 1901. The relation of sparrows to agriculture. U.S. Dep. Agri. Biol. Surv. Bull. 15:1–98.

Kantrud, H. A. 1982. *Maps of distribution and abundance of selected species of birds on uncultivated native upland grasslands and shrubsteppe in the northern Great Plains.* Biol. Serv. Prog. FWS/OBS-82/31. Washington DC: U.S. Fish and Wildlife Serv.

———, and R. L. Kologiski. 1983. Avian associations of the northern Great Plains grasslands. *J. Biogeog.* 10:331–50.

Kapustka, L. A., and J. D. DuBois. 1982. Legume distribution and nodulation in Arapaho prairie, Arthur County, Nebraska. *Prairie Nat.* 14:1–5.

Kaufmann, D. W., and E. D. Fleharty. 1974. Habitat selection by nine species of rodents in north-central Kansas. *Southwestern Nat.* 18:443–51.

Kaufmann, S. A., and J. D. Lynch, 1991. Courtship, eggs and development of the Plains topminnow in Nebraska (Actinoptergyii: Fundulidae). *Prairie Nat.* 23:41–45.

Kaul, R. 1975. *Vegetation map of Nebraska, circa 1850.* Lincoln: Univ. Neb. Cons. and Surv. Div.; rev. ed. 1993, with 32 pp. of text; also published as insert in *Nebraskaland* 71 (1): 18–19.

——. 1989. Plants. In Bleed and Flowerday 1989, pp. 127–42.

——, G. E. Kantak, and S. P. Churchill. 1988. The Niobrara River Valley, a postglacial migration corridor and refugium of forest plants and animals in the grasslands of central North America. *Bot. Rev.* 54:44–81.

Keech, C. F., and R. Bentall. 1971. Dunes on the plains. Resource Rept. 4. Lincoln: Univ. Neb. Cons. and Surv. Div.

Keeler, K. H. 1987. Survivorship and fecundity of the polycarpic perennial *Mentzelia nuda* (Loasaceae) in Nebraska Sandhills prairie. *Am. J. Bot.* 74:785–91.

——. 1991. Survivorship and recruitment in a long-lived perennial, *Ipomoea leptophylla* (Convolvulaceae). *Am. Midl. Nat.* 126:44–60.

——, A. T. Harrison, and L. Vescio. 1980. The flora and Sand Hills prairie communities of Arapaho Prairie, Arthur County, Nebraska. *Prairie Nat.* 12:65–78.

Keen, K. L. 1992. Geomorphology, recharge, and water-table fluctuations in stabilized sand dunes: The Nebraska Sand Hills, U.S.A. Ph.D. diss., Univ. of Minnesota–Minneapolis.

Kellogg, R. S. 1905. Forest belts of western Kansas and Nebraska. *U.S. For. Serv. Bull.* 66.

King, J. A. 1955. *Social behavior, social organization, and population dynamics of a black-tailed prairie dog town in the Black Hills of South Dakota.* Contr. Lab. Vert. Biol. no. 67. Ann Arbor: Univ. of Michigan.

——, ed. 1968. *Biology of* Peromyscus *(Rodentia).* Am. Soc. Mamm. Spec. Publ. no. 2

Krapu, G., ed. 1981. *The Platte River ecology study.* Northern Prairie Wildl. Res. Center, Spec. Res. Rpt. Jamestown ND: U.S. Fish and Wildlife Serv.

Kromm, D. E., and S. White, eds. 1993. *Groundwater exploitation in the High Plains.* Lawrence: Univ. Press of Kansas.

Labedz, T. E. 1989. Birds. In Bleed and Flowerday 1989, pp. 161–80.

Langemeier, R. 1984. Life in the Sandhills: Irrigated crop agriculture point of view. In *Proceedings, 1984 Water Resources Seminar Series,* pp. 89–91. Lincoln: Univ. of Nebraska.

Lanner, D. 1992. Discovery of dune dams reveals formation of Sand Hills lakes. In *Annual report,* pp. 9–12. Resource Notes 6. Lincoln: Univ. Neb. Cons. and Surv. Div.

Lawson, M. P., K. F. Dewey, and R. E. Neild. 1977. *Climatic atlas of Nebraska.* Lincoln: Univ. of Nebraska Press.

Lechleitner, R. R. 1969. *Wild mammals of Colorado: Their appearance, habits, distribution, and abundance.* Boulder CO: Pruett.

Legler, J. M. 1960. Natural history of the ornate box turtle, *Terrapene ornata ornata. Univ. Kansas Mus. Nat. Hist. Publ.* 11:527–669.

Lemen, C., and P. W. Freeman. 1986. Habitat selection and movement in communities of rodents. *Prairie Nat.* 18:129–41.

Lichty, J. A. 1960. A history of the settlement of the Nebraska Sandhills. M.A. thesis, Univ. of Wyoming–Laramie.

Lommasson, R. 1973. *Nebraska wild flowers.* Lincoln: Univ. of Nebraska Press.

Louda, S. M., M. A. Potvin, and S. K. Collinge. 1900. Predispersal seed predation, postdispersal seed predation, and competition in the recruitment of seedlings of a native thistle in Sandhills prairie. *Am. Midl. Nat.* 124:105–13.

Lugn, A. L. 1935. *The Pleistocene geology of Nebraska.* Neb. Geol. Surv. Bull. 10, 2d ser.

——. 1939. Classification of the Tertiary system in Nebraska. *Bull. Geol. Soc. Am.* 50:1245–76.

Lynch, J. D. 1985. Annotated checklist of the amphibians and reptiles of Nebraska. *Trans. Neb. Acad. Sci.* 13:33–57.

McCarraher, D. B. 1970. Some ecological relations of fairy shrimp in alkaline habitats of Nebraska. *Am. Midl. Nat.* 84:59–68.

——. 1977. *Nebraska's Sandhills lakes.* Lincoln: Nebraska Game and Parks Comm.

McClure, H. E. 1944. Mourning doves in Nebraska and the West. *Auk* 63:24–42.

——. 1966. Some observations of vertebrate fauna of the Nebraska Sandhills, 1941 through 1943. *Neb. Bird Rev.* 34:2–15.

McGregor, R. L., and T. M. Barkley, eds. 1986. *Flora of the Great Plains.* Lawrence: Univ. Press of Kansas.

McIntosh, C. B. 1975. Use and abuse of the Timber Culture Act. *Ann. Assoc. Am. Geog.* 65:355–57.

McKenzie, J. A. 1969. The courtship behaviour of the male brook stickleback *(Culaea inconstans)* (Kirkland). *Can. J. Zool.* 47:1281–86.

McMurtry, M. S., R. Craig, and G. Schildman. 1972. *Nebraska wetland survey.* Lincoln: Nebraska Game and Parks Comm.

Madsen, T. E. 1985. The status and distribution of the uncommon fishes of Nebraska. M.S. thesis, Univ. of Nebraska–Omaha.

Madson, J. 1978. Nebraska's Sand Hills: Land of long sunsets. *National Geographic* 154 (4): 493–517.

Mahoney, D. 1977. Species richness and diversity of aquatic vascular plants in Nebraska with special reference to water quality parameters. M.S. thesis, Univ. of Nebraska–Lincoln.

Marony, D. G. 1978. A stratigraphic and paleoecologic study of some late Cenozoic sediments in the central Sandhills Province of Nebraska. Ph.D. diss., Univ. of Nebraska–Lincoln.

Martin, D. L. 1984. Possible changes in the Sandhills: Ground and surface water quality and other environmental impacts. In *Proceedings, 1984 Water Resources Seminar Series,* pp. 109–20. Lincoln: Univ. of Nebraska.

Maxwell, M. H., and L. N. Brown. 1968. Ecological distribution of rodents on the High Plains of eastern Wyoming. *Southwestern Nat.* 13:143–58.

Mengel, R. M. 1970. The North American central plains as an isolating agent in bird speciation. In Dort and Jones 1970, pp. 278–340.

Mentzer, L. W. 1950. Studies of plant succession in true prairie. Ph.D. diss., Univ. of Nebraska–Lincoln.

Meserve, P. 1971. Population ecology of the prairie vole, *Microtus ochrogaster,* in the western mixed prairie of Nebraska. *Am. Midl. Nat.* 86:417–33.

Miller, S. M. 1989. Land development and use. In Bleed and Flowerday 1989, pp. 207–26.

Millstead, W. W., ed. 1967. *Lizard ecology: A symposium.* Columbia: Univ. of Missouri Press.

Morris, J., L. Morris, and L. Witt. 1972. *The fishes of Nebraska.* Lincoln: Nebraska Game and Parks Comm.

Moulton, M. P. 1972. The small playa lakes of Nebraska: Their ecology, fisheries and biological potential. In *Playa Lakes Symposium Transactions,* comp. Intern. Center for Arid and Semi-Arid Land Studies, pp. 15–23. Lubbock: Texas Tech. Univ.

——, J. R. Choate, S. J. Bissell, and R. A. Nicholson. 1981. Associations of small mammals on the central High Plains of eastern Colorado. *Southwestern Nat.* 26:53–57.

Murray, G. L. 1986. Center-pivot irrigation systems in Nebraska, 1985 [map]. Remote Sensing Center, Univ. Neb. Cons. and Surv. Div.

Nebraska Department of Agriculture. 1991. *Nebraska agricultural statistics: 1990 annual report.* Lincoln: Nebraska Crop and Livestock Reporting Service.

Nebraska Department of Economic Development. 1991. *Nebraska statistical handbook.* Lincoln NE.

Nebraska Division of Noxious Weeds. 1947. *Nebraska weeds.* Bull. 101. Lincoln: Neb. Dept. Agric. and Inspection.

Nebraska Game and Parks Comm. 1983. Nebraska rivers. *Nebraskaland* (spec. issue) 61 (1): 1–146.

——. 1985. Birds of Nebraska. *Nebraskaland* (spec. issue) 63 (1): 1–146.

——. 1987. The fish book. *Nebraskaland* (spec. issue) 65 (1): 1–132.

——. 1990. A wildflower year. *Nebraskaland* (spec. issue) 68 (1): 1–98.

——. 1993. Walk in the woods. *Nebraskaland* (spec. issue) 71 (1): 1–98.

Nebraska Water Resource Center. 1984. The Sandhills of Nebraska—yesterday, today and tomorrow. In *Proceedings, 1984 Water Resource Seminar Series,* Lincoln: Univ. of Nebraska.

Nickerson, D. 1993. Western forest. *Nebraskaland* 71 (1): 72–83.

Nielsen, E., and L. Lee. 1987. *Magnitude and costs of groundwater contamination from agricultural chemicals.* Washington DC: U.S. Dept. Agric., Econ. Res. Serv.

Nixon, E. S. 1967. A vegetational study of the Pine Ridge of northwestern Nebraska. *Southwestern Nat.* 12:134–45.

Novacek, J. M. 1989. The water and wetland resources of the Nebraska Sandhills. In *Northern Prairie Wetlands,* ed. A. van der Valk, pp. 341–84. Ames: Iowa State Univ. Press.

Nuechterlein, G. L., and R. W. Storer. 1982. The pair-formation displays of the western grebe. *Condor* 94:350–69.

Oberholser, H. C., and W. L. McAtee. 1920. *Waterfowl and their food plants in the Sandhill region of Nebraska.* Washington DC: U.S. Dept. Agr. Bull. 794.

Oliver, D. R. 1971. Life history of the Chironomidae. *Ann. Rev. Entomol.* 16:211–30.

Otte, D. 1981. *The North American grasshoppers.* Vol. 1. Cambridge MA: Harvard Univ. Press.

Pflieger, W. L. 1975. *The fishes of Missouri.* Jefferson City: Missouri Dept. of Conservation.

Platt, D. R. 1969. Natural history of the hognose snakes *Heterodon platyrhinos* and *H. nasicus.* Univ. Kansas Mus. Nat. Hist. Pub. 18:253–420.

Pool, R. J. 1914. A study of the vegetation of the Sandhills of Nebraska. *Minn. Bot. Stud.* (Univ. Minnesota–Minneapolis) 4 (3): 189–312.

Potter, J. 1984. Historical settlement patterns in the Sandhills. In *Proceedings, 1984 Water Resources Seminar Series,* pp. 16–31. Lincoln: Univ. of Nebraska.

Potvin, M. A. 1988. Seed rain in a Nebraska Sandhills prairie. *Prairie Nat.* 20:81–89.

Pound, R., and F. E. Clements. 1989. *The phytogeography of Nebraska.* Lincoln: Botanical Survey of Nebraska; rpt. 1900, Lincoln: Univ. of Nebraska Botanical Seminar.

Proctor, M., and P. Yeo. 1972. *The pollination of flowers.* New York: Taplinger.

Ramaley, F. 1939. Sand-hill vegetation of northeastern Colorado. *Ecol. Monog.* 9:1–51.

Randall, J. A. 1993. Behavioural adaptations of desert rodents (Heteromyidae). *Anim. Behav.* 45:263–87.

Ratcliffe, B. C. 1990. Prairie scarabs: The Nebraska perspective. *Am. Entomol.* 36:28–35.

——. 1991. *The scarab beetles of Nebraska.* Bull. Neb. State Mus. no. 12. Lincoln: Univ. of Nebraska.

Reichman, O. J. 1987. *Konza Prairie: A tallgrass natural history.* Lawrence: University Press of Kansas.

Reiser, J. 1986. Cedar Point Biological Station. *Neb. Lepidopterist* 2 (4): 1–3.

Reisman, H. M., and T. J. Cade. 1967. Physiological and behavioral aspects of reproduction in the brook stickleback, *Culaea inconstans. Am. Midl. Nat.* 77:257–95.

Ribble, D. O., and F. B. Samson. 1988. Microhabitat associations of small mammals in southeastern Colorado, with special emphasis on *Peromyscus. Southwestern Nat.* 32:291–303.

Riley, C. V. 1892. The yucca moth and yucca pollination. *Rept. Mo. Bot. Gdn.* 3:99–158.

Risser, P. G., E. C. Birney, H. D. Blocker, S. W. May, W. J. Parton, and J. A. Weins. 1981. *The true prairie ecosystem.* Stroudsburg PA: Hutchinson Ross.

Robins, C. E., R. M. Bailey, C. E. Bond, J. R. Brooker, E. A. Lachner, R. N. Lea, and W. B. Scott. 1991. Common and scientific names of fishes from the United States and Canada. *Am. Fisheries Soc. Spec. Pub.* 20:1–183.

Roesler, T. W., F. C. Lamphear, and M. D. Beveridge. 1968. *The economic impact of irrigated agriculture on the economy of Nebraska.* Nebraska Economic and Business Reports No. 4. Lincoln NE: Bur. of Bus. Res.

Rosche, R. C. 1982. *Birds of northwestern Nebraska and southwestern South Dakota.* Chadron NE: R. C. Rosche.

——, and P. A. Johnsgard. 1984. Birds of Lake McConaughy and the North Platte Valley (Oshkosh to Keystone). *Neb. Bird Rev.* 52:26–35; 58:84–87.

Ruffer, D. G. 1965. Sexual behaviour of the northern grasshopper mouse *(Onychomys leucogaster) Anim. Behav.* 13:447–52.

——. 1968. Agonistic behavior of the northern grasshopper mouse *(Onychomys leucogaster breviauritus). J. Mamm.* 49:481–87.

Rundquist, D. C. 1983. Wetland inventories of Nebraska's Sandhills. Resource Rept. No. 9. Lincoln: Univ. Neb. Cons. and Surv. Div.

——, S. A. Sampson, and D. E. Bussom. 1981. *Wetland atlas of the Omaha district.* Omaha: U.S. Army Corps of Engineers.

Rydberg, P. A. 1895. Flora of the Sand Hills of Nebraska. *Contrib. U.S. Nat. Herb.* 3 (3): 133–203.

Sather, J. H. 1958. Biology of the Great Plains muskrat in Nebraska. *Wildl. Monogr.* 2:1–35.

Schmidt, T. L. 1986. *Forestland resources of the Nebraska Sandhills* Ed. T. D. Wardle. Lincoln: Nebraska State Forest Service.

Schnagl, J. 1980. Seasonal variations in water chemistry and primary productivity in four alkaline lakes in the Sandhills of western Nebraska. M.S. thesis, Univ. of Nebraska–Lincoln.

Schultz, C. B., and J. C. Frye, eds. 1968. *Loess and related eolian deposits of the world.* Lincoln: Univ. of Nebraska Press.

Schwartz, C. W., and E. R. Schwartz. 1981. *The wild mammals of Missouri.* 2d ed. Columbia: Univ. of Missouri Press and Missouri Conservation Comm.

Schwilling, M. D. 1962. Species and nesting density of birds in grassland habitats near Burwell, Nebraska, in 1960. *Neb. Bird Rev.* 30:9–11.

Sharpe, R. S., and R. R. Payne. 1966. Nesting birds of the Crescent Lake National Wildlife Refuge. *Neb. Bird Rev.* 34:31–34.

Smith, H. T. U. 1965. Dune morphology and chronology in central and western Nebraska. *J. Geol.* 73:557–78.

——. 1968. Nebraska dunes compared with those of North Africa and other regions. In Schultz and Frye 1968, pp. 29–47.

Smith, R. E. 1967. National history of the prairie dog in Kansas. *Univ. Kansas Mus. Nat. Hist. Misc. Publ.* 49:1–39.

Steiger, T. L. 1930. Structure of prairie vegetation. *Ecology* 11:170–211.

Steinauer, G. 1992. Sandhills fens. *Nebraskaland* 70 (6): 17–29.

——. 1993. The Niobrara Valley forests. *Nebraskaland* 71 (1): 84–85.

Stubbendieck, J. 1989. Range management. In Bleed and Flowerday 1989, pp. 227–32.

——, and E. C. Conard. 1989. *Common legumes of the Great Plains.* Lincoln: Univ. of Nebraska Press.

——, S. L. Hatch, and K. J. Hirsch. 1986. *North American range plants.* Lincoln: Univ. of Nebraska Press.

——, J. T. Nichols, and C. H. Butterfield. 1989. *Nebraska range and pasture forbs and herbs.* Lincoln: Univ. of Nebraska Inst. Agri. and Nat. Res.

Sutherland, D. M. 1984. Vegetative key to grasses of the Sand Hills region of Nebraska. *Trans. Neb. Acad. Sci.* 12:23–60.

——, and R. B. Kaul. 1986. Nebraska plant distribution. *Trans. Neb. Acad. Sci.* 14:55–59.

——, and S. B. Rolfsmeier, 1989. An annotated list of the vascular plants of Keith County, Nebraska. *Trans. Neb. Acad. Sci.* 17:83–101.

Swinehart, J. B. 1989. Wind-blown deposits. In Bleed and Flowerday 1989, pp. 43–56.

——, and R. F. Diffendal Jr. 1989. Geology of the pre-dune strata. In Bleed and Flowerday 1989, pp. 29–42.

——, J. W. Goeke, and T. C. Winter. 1988. *Field guide to geology and hydrology of the Nebraska Sandhills.* Lincoln: Univ. Neb. Cons. and Surv. Div.

Tolstead, W. L. 1942. Vegetation of the northern part of Cherry County, Nebraska. *Ecol. Monog.* 12:255–92.

——. 1947. Woodlands in northwestern Nebraska. *Ecology* 28:180–88.

Turner, J. K., and D. C. Rundquist. 1980. *Wetlands inventory of the Omaha district: Final report.* Omaha: U.S. Army Corps of Engineers.

Twedt, C. 1974. Characteristics of sharp-tailed grouse display grounds in the Nebraska Sandhills. Ph.D. diss., Univ. of Nebraska–Lincoln.

——. 1986. Blowout plants. *Nebraskaland* 64 (2): 49–50.

Udvardy, M. D. F. 1957. An evaluation of quantitative studies in birds. *Cold Spring Harbor Symp. Quant. Biol.* 11:301–11.

U.S. Department of Agriculture. 1991. *Nebraska agricultural statistics, 1990.* Lincoln NE.

U.S. Geological Survey. 1984. *National water summary.* Reston VA.

——. 1991. *Water level changes in the High Plains aquifer: Predevelopment to 1990.* Lincoln NE.

Vance, F. R., J. R. Jowsey, and J. S. McLean. 1984. *Wildflowers of the northern Great Plains*. Minneapolis: Univ. of Minnesota Press.

Walker, J. A. 1966. Summer habitat of sharp-tailed grouse, *Pedioecetes phasianellus* (Linnaeus), in the Nebraska Sandhills. M.S. thesis, Ft. Hays State College, Kansas.

Warren, A. 1976. Morphology and sediments of the Nebraska Sandhills in relation to Pleistocene winds and the development of aeolian bedforms. *J. Geol.* 84: 685–700.

——. 1983. Late Quaternary vegetation of the Great Plains. *Trans. Neb. Acad. Sci.* 11:83–89.

Watts, W. A., and H. E. Wright Jr. 1966. Late Wisconsin pollen and seed analysis from the Nebraska Sandhills. *Ecology* 47:202–10.

Weaver, J. E. 1954. *North American prairie*. Lincoln: Johnson.

——. 1965. *Native vegetation of Nebraska*. Lincoln: Univ. of Nebraska Press.

——. 1968. *Prairie plants and their environment: A fifty-year study in the Midwest*. Lincoln: Univ. of Nebraska Press.

——, and F. W. Albertson. 1956. *Grasslands of the Great Plains*. Lincoln: Johnson.

Webster, D. B. 1962. A function of the enlarged middle-ear cavities of the kangaroo rat, *Dipodomys*. *Physiol. Zool.* 32:248–55.

Weller, S. G. 1985. The life history of *Lithospermum carolinensis,* a long-lived herbaceous sand dune species. *Ecol. Monog.* 55:49–67.

Wells, P. V. 1970. Postglacial vegetational history of the Great Plains. *Science* 167:1574–82.

Wiens, J. A. 1973. Pattern and process in grassland bird communities. *Ecol. Monog.* 43:237–70.

Williams, J. H., and D. Murfield. 1977. *Agricultural atlas of Nebraska*. Lincoln: Univ. of Nebraska Press.

Winn, H. E. 1960. Biology of the brook stickleback *Eucalia inconstans* (Kirtland). *Am. Midl. Nat.* 63:424–38.

Wolcott, R. H. 1909. An analysis of Nebraska's bird fauna. *Proc. Neb. Ornith. Union* 4:22–55.

Wooton, R. J. 1976. *The biology of the sticklebacks*. New York: Academic Press.

Wright, H. E. 1970. Vegetational history of the Great Plains. In Dort and Jones 1970, pp. 185–202.

——, J. C. Almendinger, and J. Grüger. 1985. Pollen diagram from the Nebraska Sandhills and the age of the dunes. *Quaternary Res.* 24:115–20.

Zimmer, J. T. 1913. Birds of the Thomas County Forest Reserve. *Proc. Neb. Ornith. Union* 5:51–104.

Zimmerman, J. L. 1990. *Cheyenne Bottoms: Wetland in jeopardy*. Lawrence: Univ. Press of Kansas.

——. 1993. *The birds of Konza: Avian ecology of the tallgrass prairie*. Lawrence: Univ. Press of Kansas.

Zwingle, E. 1993. Ogallala aquifer: Wellspring of the High Plains. *National Geographic* 183 (3): 80–109.

Subject Index

This index is largely limited to topics relevant to the Sandhills and Nebraska, and especially to geographic locations and important technical biological or geological terms that are mentioned in the text. Materials included in the appendices have not been incorporated into either this index or the following one.

Platte River Valley, 3, 34–44, 105, 107, 135, 137, 151, 153
Pleistocene epoch, 6, 15, 21, 35
Pliocene epoch, 6, 36, 37, 38
Poison Ivy Quarry, 6
Polk County, 151

Quaternary period, 38

Rock County, 13, 16, 23, 138, 142–45, 153
Rosebud Formation, 16, 19

Sand draws, 72, 77
Sand sheets, 10, 13
Scotts Bluff National Monument, 4, 25, 26, 28, 33
Seminole Dam (wy), 36
Sheridan County, 8–10, 27, 120, 131, 138
Sioux County, 27
Slip face, 11
Snake River, 105, 106
South Loup River, 106
South Platte River, 106
Succession patterns in prairie, 48, 51, 75
Suture zones, 78

Symbiosis, 80
Sympatric distributions, 20

Tertiary period, 21, 35
Thedford, 61
Thomas County, 138, 144
Thurston County, 154
Timber Culture Act, 136

Valentine, 15, 16, 18
Valentine Formation, 16, 19, 21
Valentine National Wildlife Refuge, 58
Valley (city), 142
Valley County, 58, 138, 145

Walthill, 154
Warm-season (C-4) grasses, 94
Wheeler County, 13, 138, 142–45, 154
White River, 25, 106
White River Group, 4
Wildcat Hills, 27
Wisconsinian glaciation, 6, 9

York County, 151

Index to Plants and Animals

Stiff sunflower, 70, 72
Stonewort, 119, 120
Striped skunk, 26
Swainson's hawk, 26, 38, 58, 62
Swainson's thrush, 29
Swamp lousewort, 119
Swamp sparrow, 19, 22, 119
Sweet clover, 73
Swift fox, 50
Switchgrass, 48, 70, 75, 94, 118

Thirteen-lined ground squirrel, 19, 31, 32, 50
Three-awn grasses, 72, 78
Three-square, 121
Townsend's solitaire, 26, 29
Tree swallow, 18
Trumpeter swan, 131–32
Tumbleweeds, 73
Turkey vulture, 26, 29

Upland sandpiper, 26, 29, 38, 41, 50, 52, 56, 82, 83, 102, 103

Vesper sparrow, 18, 22, 26, 32, 41, 42, 83
Violet-green swallow, 26, 29
Virginia rail, 42

Wandering gartersnake, 30
Warbling vireo, 19, 28
Water crowfoot, 119
Water hemlock, 73
Water horsetail, 119
Waterlily, 73
Water milfoil, 118
Water plantain, 105
Western chorus frog, 26
Western flycatcher, 29
Western grebe, 128–31
Western harvest mouse, 26, 31, 50
Western hognose snake, 23, 26
Western kingbird, 38, 40, 41

Western meadowlark, 19, 22, 26, 32, 40–42, 50, 59, 62, 83
Western ragweed, 71, 78
Western redcedar, 27
Western red lily, 119
Western tanager, 20, 29
Western wheatgrass, 28, 29, 70, 75, 94, 95, 118
Western wood-pewee, 20, 29
Whip-poor-will, 18, 19, 23
White beardtongue, 72
White-breasted nuthatch, 19, 26, 52
White-footed mouse, 51
White sage, 70
White-tailed deer, 135
White-tailed jackrabbit, 50, 58, 60
White-throated swift, 26, 29
White-whiskered grasshopper, 95
Wigeon grass, 119
Wild alfalfa, 30, 71
Wild barley, 72
Wild lettuce, 73
Wild plum, 22, 27
Wild turkey, 28, 40
Willow flycatcher, 41
Willowherb, 105
Willows, 14, 22, 28, 73
Winged pigweed, 73
Wood duck, 19, 28
Woodhouse's toad, 26
Wood thrush, 18, 19, 23

Yellow-billed cuckoo, 28, 38, 52
Yellow-breasted chat, 19, 22
Yellow-headed blackbird, 19, 104, 119
Yellow-rumped warbler, 26, 29
Yellow-throated vireo, 19
Yellow warbler, 26, 28, 40, 41
Yellow water lily, 118, 119
Yellow wood sorrel, 73
Yucca. *See* Small soapweed
Yucca moth, 79–80

Zebras, 6